Solutions Manual to Accompany

An Introduction to Numerical Methods and Analysis

Solutions Manual to Accompany

An Introduction to Numerical Methods and Analysis

THIRD EDITION

James F. Epperson

Mathematical Reviews, American Mathematical Society

This third edition first published 2021
© 2021 John Wiley & Sons, Inc.

Edition History
John Wiley and Sons, Inc. (2e, 2014)

The right of James F. Epperson to be identified as the author of this work has been asserted in accordance with law.

Registered Office
John Wiley & Sons, Inc., 111 River Street, Hoboken, NJ 07030, USA

Editorial Office
111 River Street, Hoboken, NJ 07030, USA

For details of our global editorial offices, customer services, and more information about Wiley products visit us at www.wiley.com.

Wiley also publishes its books in a variety of electronic formats and by print-on-demand. Some content that appears in standard print versions of this book may not be available in other formats.

Library of Congress Cataloging-in-Publication Data Applied for
ISBN: 9781119604532

Cover design by Wiley

Set in 9/11pt NimbusRomNo9L by Straive, Chennai, India

10 9 8 7 6 5 4 3 2 1

CONTENTS

Preface to the Solutions Manual for the Third Edition

This manual is written for instructors, not students. It includes worked solutions for many (roughly 75%) of the problems in the text. For the computational exercises I have given the output generated by my program, or sometimes a program listing. Most of the programming was done in MATLAB, some in FORTRAN. (The author is well aware that FORTRAN is archaic, but there is a lot of "legacy code" in FORTRAN, and the author believes there is value in learning a new language, even an archaic one.) When the text has a series of exercises that are obviously similar and have similar solutions, then sometimes only one of these problems has a worked solution included. When computational results are asked for a series of similar functions or problems, only a subset of solutions are reported, largely for the sake of brevity. Some exercises that simply ask the student to perform a straight-forward computation are skipped. Exercises that repeat the same computation but with a different method are also often skipped, as are exercises that ask the student to "verify" a straight-forward computation.

Some of the exercises were designed to be open-ended and almost "essay-like." For these exercises, the only solution typically provided is a short hint or brief outline of the kind of discussion anticipated by the author.

In many exercises the student needs to construct an upper bound on a derivative of some function in order to determine how small a parameter has to be to achieve a desired level of accuracy. For many of the solutions this was done using a computer algebra package and the details are not given.

Students who acquire a copy of this manual in order to obtain worked solutions to homework problems should be aware that none of the solutions are given in enough detail to earn full credit from an instructor.

The author freely admits the potential for error in any of these solutions, especially since many of the exercises were modified after the final version of the text was submitted to the publisher and because the ordering of the exercises was changed between editions. While we tried to make all the appropriate corrections, the possibility of error is still present, and undoubtedly the author's responsibility.

Because much of the manual was constructed by doing "copy-and-paste" from the files for the text, the enumeration of many tables and figures will be different. I have tried to note what the number is in the text, but certainly may have missed some instances.

Suggestions for new exercises and corrections to these solutions are very welcome. Contact the author at jfe@ams.org or jfepperson@gmail.com.

Differences from the text   The text itself went through a copy-editing process after this manual was completed. As was to be expected, the wording of several problems was slightly changed. None of these changes should affect the problem in terms of what is expected of students; the vast majority of the changes were to replace "previous problem" (a bad habit of mine) with "Problem X.Y" (which I should have done on my own, in the first place). Some punctuation was also changed. The point of adding this note is to explain the textual differences which might be noticed between the text and this manual. If something needs clarification, please contact me at the above email.

CHAPTER 1

INTRODUCTORY CONCEPTS AND CALCULUS REVIEW

1.1 BASIC TOOLS OF CALCULUS

Exercises:

1. Show that the third-order Taylor polynomial for $f(x) = (x+1)^{-1}$, about $x_0 = 0$, is

$$p_3(x) = 1 - x + x^2 - x^3.$$

Solution: We have $f(0) = 1$ and

$$f'(x) = -\frac{1}{(x+1)^2}, \quad f''(x) = \frac{2}{(x+1)^3}, \quad f'''(x) = -\frac{6}{(x+1)^4},$$

so that $f'(0) = -1$, $f''(0) = 2$, $f''' = -6$. Therefore,

$$
\begin{aligned}
p_3(x) &= f(0) + xf'(0) + \frac{1}{2}x^2 f''(0) + \frac{1}{6}x^3 f'''(x) \\
&= 1 + x(-1) + \frac{1}{2}x^2(2) + \frac{1}{6}x^3(-6) \\
&= 1 - x + x^2 - x^3.
\end{aligned}
$$

2. What is the third-order Taylor polynomial for $f(x) = \sqrt{x+1}$, about $x_0 = 0$?

Solutions Manual to Accompany An Introduction to Numerical Methods and Analysis, Third Edition.
James F. Epperson.
© 2021 John Wiley & Sons, Inc. Published 2021 by John Wiley & Sons, Inc.

Solution: We have $f(x_0) = 1$ and

$$f'(x) = \frac{1}{2(x+1)^{1/2}}, \quad f''(x) = -\frac{1}{4(x+1)^{3/2}}, \quad f'''(x) = \frac{3}{8(x+1)^{5/2}},$$

so that $f'(0) = 1/2$, $f''(0) = -1/4$, $f''' = 3/8$. Therefore

$$
\begin{aligned}
p_3(x) &= f(0) + xf'(0) + \frac{1}{2}x^2 f''(0) + \frac{1}{6}x^3 f'''(x) \\
&= 1 + x(1/2) + \frac{1}{2}x^2(-1/4) + \frac{1}{6}x^3(3/8) \\
&= 1 - (1/2)x - (1/8)x^2 + (1/16)x^3.
\end{aligned}
$$

3. What is the sixth-order Taylor polynomial for $f(x) = \sqrt{1 + x^2}$, using $x_0 = 0$? Hint: Consider the previous problem.

4. Given that

$$R(x) = \frac{|x|^6}{6!}e^{\xi}$$

for $x \in [-1, 1]$, where ξ is between x and 0, find an upper bound for $|R|$, valid for all $x \in [-1, 1]$, that is independent of x and ξ.

5. Repeat the above, but this time require that the upper bound be valid only for all $x \in [-\frac{1}{2}, \frac{1}{2}]$.

   Solution: The only significant difference is the introduction of a factor of 2^6 in the denominator:

   $$|R(x)| \leq \frac{\sqrt{e}}{2^6 \times 720} = 3.6 \times 10^{-5}.$$

6. Given that

$$R(x) = \frac{|x|^4}{4!}\left(\frac{-1}{1+\xi}\right)$$

for $x \in [-\frac{1}{2}, \frac{1}{2}]$, where ξ is between x and 0, find an upper bound for $|R|$, valid for all $x \in [-\frac{1}{2}, \frac{1}{2}]$, that is independent of x and ξ.

7. Use a Taylor polynomial to find an approximate value for $\sqrt{e}$ that is accurate to within 10^{-3}.

   Solution: There are two ways to do this. We can approximate $f(x) = e^x$ and use $x = 1/2$, or we can approximate $g(x) = \sqrt{x}$ and use $x = e$. In addition, we can be conventional and take $x_0 = 0$, or we can take $x_0 \neq 0$ in order to speed convergence.

   The most straightforward approach (in my opinion) is to use a Taylor polynomial for e^x about $x_0 = 0$. The remainder after k terms is

   $$R_k(x) = \frac{x^{k+1}}{(k+1)!}e^{\xi}.$$

   We quickly have that

   $$|R_k(x)| \leq \frac{e^{1/2}}{2^{k+1}(k+1)!}$$

and a little playing with a calculator shows that

$$|R_3(x)| \leq \frac{e^{1/2}}{16 \times 24} = 0.0043$$

but

$$|R_4(x)| \leq \frac{e^{1/2}}{32 \times 120} = 4.3 \times 10^{-4}.$$

So we would use

$$e^{1/2} \approx 1 + \frac{1}{2} + \frac{1}{2}\left(\frac{1}{2}\right)^2 + \frac{1}{6}\left(\frac{1}{2}\right)^3 + \frac{1}{24}\left(\frac{1}{2}\right)^4 = 1.6484375.$$

To fourteen digits, $\sqrt{e} = 1.64872127070013$, and the error is 2.84×10^{-4}, much smaller than required.

8. What is the fourth-order Taylor polynomial for $f(x) = 1/(x+1)$, about $x_0 = 0$?

   Solution: We have $f(0) = 1$ and

   $$f'(x) = -\frac{1}{(x+1)^2}, \quad f''(x) = \frac{2}{(x+1)^3}, \quad f'''(x) = -\frac{6}{(x+1)^4}, \quad f''''(x) = \frac{24}{(x+1)^5},$$

   so that $f'(0) = -1$, $f''(0) = 2$, $f''' = -6$, $f''''(0) = 24$. Thus,

   $$p_4(x) = 1 + x(-1) + \frac{1}{2}x^2(2) + \frac{1}{6}x^3(-6) + \frac{1}{24}x^4(24) = 1 - x + x^2 - x^3 + x^4.$$

9. What is the fourth-order Taylor polynomial for $f(x) = 1/x$, about $x_0 = 1$?

10. Find the Taylor polynomial of third-order for $\sin x$, using:

    (a) $x_0 = \pi/6$.

    Solution: We have

    $$f(x_0) = \frac{1}{2}, \quad f'(x_0) = \frac{\sqrt{3}}{2}, \quad f''(x_0) = -\frac{1}{2}, \quad f'''(x_0) = -\frac{\sqrt{3}}{2},$$

    so

    $$p_3(x) = \frac{1}{2} + \frac{\sqrt{3}}{2}\left(x - \frac{\pi}{6}\right) - \frac{1}{4}\left(x - \frac{\pi}{6}\right)^2 - \frac{\sqrt{3}}{12}\left(x - \frac{\pi}{6}\right)^3;$$

    (b) $x_0 = \pi/4$;

    (c) $x_0 = \pi/2$.

11. For each function below construct the third-order Taylor polynomial approximation, using $x_0 = 0$, and then estimate the error by computing an upper bound on the remainder, over the given interval.

    (a) $f(x) = e^{-x}$, $x \in [0, 1]$;

    (b) $f(x) = \ln(1 + x)$, $x \in [-1, 1]$;

    (c) $f(x) = \sin x$, $x \in [0, \pi]$;

    (d) $f(x) = \ln(1 + x)$, $x \in [-1/2, 1/2]$;

(e) $f(x) = 1/(x+1)$, $x \in [-1/2, 1/2]$.

Solution:

(a) The polynomial is
$$p_3(x) = 1 - x + \frac{1}{2}x^2 - \frac{1}{6}x^3,$$
with remainder
$$R_3(x) = \frac{1}{24}x^4 e^{-\xi}.$$
This can be bounded above, for all $x \in [0, 1]$, by
$$|R_3(x)| \leq \frac{1}{24}e$$

(b) The polynomial is
$$p_3(x) = x - \frac{1}{2}x^2 + \frac{1}{3}x^3,$$
with remainder
$$R_3(x) = \frac{1}{4}x^4 \frac{1}{(1+\xi)^4}.$$
We *can't* bound this for all $x \in [-1, 1]$, because of the potential division by zero.

(c) The polynomial is
$$p_3(x) = x - \frac{1}{6}x^3,$$
with remainder
$$R_3(x) = \frac{1}{120}x^5 \cos \xi.$$
This can be bounded above, for all $x \in [0, \pi]$, by
$$|R_3(x)| \leq \frac{\pi^5}{120}.$$

(d) The polynomial is the same as in (b), of course,
$$p_3(x) = x - \frac{1}{2}x^2 + \frac{1}{3}x^3,$$
with remainder
$$R_3(x) = \frac{1}{4}x^4 \frac{1}{(1+\xi)^4}.$$
For all $x \in [-1/2, 1/2]$ this can be bounded by
$$R_3(x) \leq \frac{1}{4}(1/2^4)\frac{1}{(1-(1/2))^4} = \frac{1}{4}.$$

(e) The polynomial is
$$p_3(x) = 1 - x + x^2 - x^3,$$
with remainder
$$R_3(x) = x^4 \frac{1}{(1+\xi)^5}.$$

This can be bounded above, for all $x \in [-1/2, 1/2]$, by

$$|R_3(x)| \le (1/2)^4 \frac{1}{(1-1/2)^5} = 2.$$

Obviously, this is not an especially good approximation.

12. Construct a Taylor polynomial approximation that is accurate to within 10^{-3}, over the indicated interval, for each of the following functions, using $x_0 = 0$.

(a) $f(x) = \sin x$, $x \in [0, \pi]$;

(b) $f(x) = e^{-x}$, $x \in [0, 1]$;

(c) $f(x) = \ln(1 + x)$, $x \in [-1/2, 1/2]$;

(d) $f(x) = 1/(x + 1)$, $x \in [-1/2, 1/2]$;

(e) $f(x) = \ln(1 + x)$, $x \in [-1, 1]$.

Solution:

(a) The remainder here is

$$R_n(x) = \frac{(-1)^{n+1}}{(2n+1)!} x^{2n+1} \cos c,$$

for $c \in [0, \pi]$. Therefore, we have

$$|R_n(x)| \le \frac{1}{(2n+1)!} |\pi|^{2n+1} \le \frac{\pi^{2n+1}}{(2n+1)!}.$$

Simple manipulations with a calculator then show that

$$\max_{x \in [0, \pi]} |R_6(x)| \le 0.4663028067 \times 10^{-3}$$

but

$$\max_{x \in [0, \pi]} |R_5(x)| \le 0.7370430958 \times 10^{-2}.$$

Therefore the desired Taylor polynomial is

$$p_{11}(x) = 1 - x + \frac{1}{6}x^3 - \frac{1}{120}x^5 - \frac{1}{7!}x^7 + \frac{1}{9!}x^9 + \frac{1}{11!}x^{11}.$$

(b) The remainder here is

$$R_n(x) = \frac{(-1)^{n+1}}{(n+1)!} x^{n+1} e^{-c},$$

for $c \in [0, 1]$. Therefore, we have

$$|R_n(x)| \le \frac{1}{(n+1)!} |x|^{n+1} \le \frac{1}{(n+1)!}.$$

Simple manipulations with a calculator then show that

$$\max_{x \in [0,1]} |R_6(x)| \le 0.0001984126984$$

but

$$\max_{x \in [0,1]} |R_5(x)| \le 0.1388888889 \times 10^{-2}$$

Therefore the desired Taylor polynomial is

$$p_6(x) = 1 - x + \frac{1}{2}x^2 - \frac{1}{6}x^3 + \frac{1}{24}x^4 - \frac{1}{120}x^5 + \frac{1}{720}x^6.$$

(c) $f(x) = \ln(1+x)$, $x \in [0, 3/4]$.

(d) **Solution:** The remainder is now

$$|R_n(x)| \le \frac{(1/2)^{n+1}}{(n+1)},$$

and $n = 8$ makes the error small enough.

(e) $f(x) = \ln(1+x)$, $x \in [0, 1/2]$.

13. Repeat the above, this time with a desired accuracy of 10^{-6}.

14. Since

$$\frac{\pi}{4} = \arctan 1,$$

we can estimate π by estimating $\arctan 1$. How many terms are needed in the Gregory series for the arctangent to approximate π to 100 decimal places? 1,000? Hint: Use the error term in the Gregory series to predict when the error gets sufficiently small.

Solution: The remainder in the Gregory series approximation is

$$R_n(x) = (-1)^{n+1} \int_0^x \frac{t^{2n+2}}{1+t^2} dt,$$

so to get 100 decimal places of accuracy for $x = 1$, we require

$$|R_n(1)| = \left| \int_0^1 \frac{t^{2n+2}}{1+t^2} dt \right| \le \int_0^1 t^{2n+2} dt = \frac{1}{2n+3} \le 10^{-100},$$

thus, we have to take $n \ge (10^{100} - 3)/2$ terms. For 1,000 places of accuracy we therefore need $n \ge (10^{1000} - 3)/2$ terms.

Obviously, this is not the best procedure for computing many digits of π!

15. Elementary trigonometry can be used to show that

$$\arctan(1/239) = 4 \arctan(1/5) - \arctan(1).$$

This formula was developed in 1706 by the English astronomer John Machin. Use this to develop a more efficient algorithm for computing π. How many terms are needed to get 100 digits of accuracy with this form? How many terms are needed to get 1,000 digits? Historical note: Until 1961, this was the basis for the most commonly used method for computing π to high accuracy.

Solution: We now have two Gregory series, thus complicating the problem a bit. We have

$$\pi = 4 \arctan(1) = 16 \arctan(1/5) - 4 \arctan(1/239).$$

Define $p_{m,n} \approx \pi$ as the approximation generated by using an m term Gregory series to approximate $\arctan(1/5)$ and an n term Gregory series for $\arctan(1/239)$. Then we have

$$p_{m,n} - \pi = 16 R_m(1/5) - 4 R_n(1/239),$$

where R_k is the remainder in the Gregory series. Therefore,

$$|p_{m,n} - \pi| \leq \left| 16(-1)^{m+1} \int_0^{1/5} \frac{t^{2m+2}}{1+t^2} dt - 4(-1)^{n+1} \int_0^{1/239} \frac{t^{2n+2}}{1+t^2} dt \right|$$

$$\leq \frac{16}{(2m+3)5^{2m+3}} + \frac{4}{(2n+3)239^{2n+3}}.$$

To finish the problem we have to apportion the error between the two series, which introduces some arbitrariness into the problem. If we require that they be equally accurate, then we have that

$$\frac{16}{(2m+3)5^{2m+3}} \leq \epsilon$$

and

$$\frac{4}{(2n+3)239^{2n+3}} \leq \epsilon.$$

Using properties of logarithms, these become

$$\log(2m+3) + (2m+3)\log 5 \geq \log 16 - \log \epsilon$$

and

$$\log(2n+3) + (2n+3)\log 239 \geq \log 4 - \log \epsilon.$$

For $\epsilon = (1/2) \times 10^{-100}$, these are satisfied for $m = 70$, $n = 20$. For $\epsilon = (1/2) \times 10^{-1000}$, we get $m = 712$, $n = 209$. Changing the apportionment of the error doesn't change the results by much at all.

16. In 1896, a variation on Machin's formula was found:

$$\arctan(1/239) = \arctan(1) - 6 \arctan(1/8) - 2 \arctan(1/57),$$

and this began to be used in 1961 to compute π to high accuracy. How many terms are needed when using this expansion to get 100 digits of π? 1,000 digits?

Solution: We now have three series to work with, which complicates matters only slightly more compared to the previous problem. If we define $p_{k,m,n} \approx \pi$ based on

$$\pi = 4 \arctan(1) = 24 \arctan(1/8) + 8 \arctan(1/57) + 4 \arctan(1/239),$$

taking k terms in the series for $\arctan(1/8)$, m terms in the series for $\arctan(1/57)$, and n terms in the series for $\arctan(1/239)$, then we are led to the inequalities

$$\log(2k+3) + (2k+3)\log 8 \geq \log 24 - \log \epsilon,$$

$$\log(2m+3) + (2m+3)\log 57 \geq \log 8 - \log \epsilon,$$

and

$$\log(2n+3) + (2n+3)\log 239 \geq \log 4 - \log \epsilon.$$

For $\epsilon = (1/3) \times 10^{-100}$, we get $k = 54$, $m = 27$, and $n = 19$; for $\epsilon = (1/3) \times 10^{-1000}$ we get $k = 552$, $m = 283$, and $n = 209$.

Note: In both of these problems a slightly more involved treatment of the error might lead to fewer terms being required.

17. What is the Taylor polynomial of order 3 for $f(x) = x^4 + 1$, using $x_0 = 0$?

    Solution: This is very direct:

    $$f'(x) = 4x^3, \quad f''(x) = 12x^2, \quad f'''(x) = 24x,$$

    so that

    $$p_3(x) = 1 + x(0) + \frac{1}{2}x^2(0) + \frac{1}{6}x^3(0) = 1.$$

18. What is the Taylor polynomial of order 4 for $f(x) = x^4 + 1$, using $x_0 = 0$? Simplify as much as possible.

19. What is the Taylor polynomial of order 2 for $f(x) = x^3 + x$, using $x_0 = 1$?

20. What is the Taylor polynomial of order 3 for $f(x) = x^3 + x$, using $x_0 = 1$? Simplify as much as possible.

    Solution: We note that $f'''(1) = 6$, so we have (using the solution from the previous problem)

    $$p_4(x) = 3x^2 - 2x + 1 + \frac{1}{6}(x-1)^3(6) = x^3 + x.$$

    The polynomial is its own Taylor polynomial.

21. Let $p(x)$ be an arbitrary polynomial of degree less than or equal to n. What is its Taylor polynomial of degree n, about an arbitrary x_0?

22. The Fresnel integrals are defined as

    $$C(x) = \int_0^x \cos(\pi t^2/2)dt,$$

    and

    $$S(x) = \int_0^x \sin(\pi t^2/2)dt.$$

    Use Taylor expansions to find approximations to $C(x)$ and $S(x)$ that are 10^{-4} accurate for all x with $|x| \leq \frac{1}{2}$. Hint: Substitute $x = \pi t^2/2$ into the Taylor expansions for the cosine and sine.

    Solution: We will show the work for the case of $S(x)$, only. We have

    $$S(x) = \int_0^x \sin(\pi t^2/2)dt = \int_0^x p_n(t^2)dt + \int_0^x R_n(t^2)dt.$$

    Looking more carefully at the remainder term, we see that it is given by

    $$r_n(x) = \pm \int_0^x \frac{t^{2(2n+3)}}{(2n+3)!} \cos \xi dt.$$

Therefore,

$$|r_n(x)| \le \int_0^{1/2} \frac{t^{2(2n+3)}}{(2n+3)!} dt = \frac{(1/2)^{4n+7}}{(4n+7)(2n+3)!}.$$

A little effort with a calculator shows that this is less than 10^{-4} for $n \ge 1$; therefore the polynomial is

$$p(x) = \int_0^x (t^2 - (1/6)t^6) dt = -\frac{x^7}{42} + \frac{x^3}{3}.$$

23. Use the Integral Mean Value Theorem to show that the "pointwise" form (1.3) of the Taylor remainder (usually called the *Lagrange* form) follows from the "integral" form (1.2) (usually called the *Cauchy* form).

24. For each function in Problem 11, use the Mean Value Theorem to find a value M such that

$$|f(x_1) - f(x_2)| \le M|x_1 - x_2|$$

is valid for all x_1, x_2 in the interval used in Problem 11.

Solution: This amounts to finding an upper bound on $|f'|$ over the interval given. The answers are as given below.

(a) $f(x) = e^{-x}$, $x \in [0, 1]$; $M \le 1$.

(b) $f(x) = \ln(1+x)$, $x \in [-1, 1]$; M is unbounded, since $f'(x) = 1/(1+x)$ and $x = -1$ is possible.

(c) $f(x) = \sin x$, $x \in [0, \pi]$; $M \le 1$.

(d) $f(x) = \ln(1+x)$, $x \in [-1/2, 1/2]$; $M \le 2$.

(e) $f(x) = 1/(x+1)$, $x \in [-1/2, 1/2]$. $M \le 4$.

25. A function is called *monotone* on an interval if its derivative is strictly positive or strictly negative on the interval. Suppose f is continuous and monotone on the interval $[a, b]$, and $f(a)f(b) < 0$; prove that there is exactly one value $\alpha \in [a, b]$ such that $f(\alpha) = 0$.

Solution: Since f is continuous on the interval $[a, b]$ and $f(a)f(b) < 0$, the Intermediate Value Theorem guarantees that there is a point c where $f(c) = 0$, i.e., there is at least one root. Suppose now that there exists a second root, γ. Then $f(c) = f(\gamma) = 0$. By the Mean Value Theorem, then, there is a point ξ between c and γ such that

$$f'(\xi) = \frac{f(\gamma) - f(c)}{\gamma - c} = 0.$$

But this violates the hypothesis that f is monotone, since a monotone function must have a derivative that is strictly positive or strictly negative. Thus we have a contradiction, thus there cannot exist the second root.

A very acceptable argument can be made by appealing to a graph of the function.

26. Finish the proof of the Integral Mean Value Theorem (Theorem 1.5) by writing up the argument in the case that g is negative.

Solution: All that is required is to observe that if g is negative, then we have

$$\int_a^b g(t)f(t)dt \le \int_a^b g(t)f_m dt = f_m \int_a^b g(t)dt,$$

and

$$\int_a^b g(t)f(t)dt \ge \int_a^b g(t)f_M dt = f_M \int_a^b g(t)dt.$$

The proof is completed as in the text.

27. Prove Theorem 1.6, providing all details.

28. Let $c_k > 0$, be given, $1 \le k \le n$, and let $x_k \in [a, b]$, $1 \le k \le n$. Then, use the Discrete Average Value Theorem to prove that, for any function $f \in C([a, b])$,

$$\frac{\sum_{k=1}^n c_k f(x_k)}{\sum_{k=1}^n c_k} = f(\xi),$$

for some $\xi \in [a, b]$.

Solution: We can't apply the Discrete Average Value Theorem to the problem as it is posed originally, so we have to manipulate a bit. Define

$$\gamma_j = \frac{c_j}{\sum_{k=1}^n c_k};$$

then

$$\sum_{j=1}^n \gamma_j = 1$$

and now we can apply the Discrete Average Value Theorem to finish the problem.

29. Discuss, in your own words, whether or not the following statement is true: "The Taylor polynomial of degree n is the best polynomial approximation of degree n to the given function near the point x_0."

1.2   ERROR, APPROXIMATE EQUALITY, AND ASYMPTOTIC ORDER NOTATION

Exercises:

1. Use Taylor's Theorem to show that $e^x = 1 + x + \mathcal{O}(x^2)$ for x sufficiently small.

2. Use Taylor's Theorem to show that $\frac{1 - \cos x}{x} = \frac{1}{2}x + \mathcal{O}(x^3)$ for x sufficiently small.

   Solution: We can expand the cosine in a Taylor series as

   $$\cos x = 1 - \frac{1}{2}x^2 + \frac{1}{24}x^4 \cos \xi.$$

If we substitute this into $(1 - \cos x)/x$ and simplify, we get

$$\frac{1 - \cos x}{x} = \frac{1}{2}x - \frac{1}{24}x^3 \cos \xi,$$

so that we have

$$\left| \frac{1 - \cos x}{x} - \frac{1}{2}x \right| = \left| \frac{1}{24}x^3 \cos \xi \right| \leq \frac{1}{24}|x^3| = C|x^3|$$

where $C = 1/24$. Therefore, $\frac{1-\cos x}{x} = \frac{1}{2}x + \mathcal{O}(x^3)$.

3. Use Taylor's Theorem to show that

$$\sqrt{1 + x} = 1 + \frac{1}{2}x + \mathcal{O}(x^2)$$

for x sufficiently small.

Solution: We have, from Taylor's Theorem, with $x_0 = 0$,

$$\sqrt{1 + x} = 1 + \frac{1}{2}x - \frac{1}{8}x^2(1 + \xi)^{-3/2},$$

for some ξ between 0 and x. Since

$$\left| \frac{1}{8}x^2(1 + \xi)^{-3/2} \right| \leq C|x^2|$$

for all x sufficiently small, the result follows. For example, we have

$$\left| \frac{1}{8}x^2(1 + \xi)^{-3/2} \right| \leq \frac{1}{8} \times 2\sqrt{2}|x^2|$$

for all $x \in [-1/2, 1/2]$.

4. Use Taylor's Theorem to show that

$$(1 + x)^{-1} = 1 - x + x^2 + \mathcal{O}(x^3)$$

for x sufficiently small.

Solution: This time, Taylor's Theorem gives us that

$$(1 + x)^{-1} = 1 - x + x^2 - x^3/(1 + \xi)^4$$

for some ξ between 0 and x. Thus, for all x such that $|x| \leq m$,

$$\left| (1 + x)^{-1} - (1 - x + x^2) \right| = \left| x^3/(1 + \xi)^4 \right| \leq |x|^3/(1 - m)^4 = C|x|^3,$$

where $C = 1/(1 - m)^4$.

5. Show that

$$\sin x = x + \mathcal{O}(x^3).$$

6. Recall the summation formula

$$1 + r + r^2 + r^3 + \cdots + r^n = \sum_{k=0}^{n} r^k = \frac{1 - r^{n+1}}{1 - r}.$$

Use this to prove that

$$\sum_{k=0}^{n} r^k = \frac{1}{1-r} + \mathcal{O}(r^{n+1}).$$

Hint: What is the *definition* of the $\mathcal{O}$ notation?

7. Use the above result to show that 10 terms ($k = 9$) are all that is needed to compute

$$S = \sum_{k=0}^{\infty} e^{-k}$$

to within 10^{-4} absolute accuracy.

Solution: The remainder in the 9 term partial sum is

$$|R_9| = \left| \frac{e^{-10}}{1-e^{-1}} \right| = 0.000071822 < 10^{-4}.$$

8. Recall the summation formula

$$\sum_{k=1}^{n} k = \frac{n(n+1)}{2}.$$

Use this to show that

$$\sum_{k=1}^{n} k = \frac{1}{2}n^2 + \mathcal{O}(n).$$

9. State and prove the version of Theorem 1.7 which deals with relationships of the form $x = x_n + \mathcal{O}(\beta(n))$.

Solution: The theorem statement might be something like the following:

Theorem: *Let* $x = x_n + \mathcal{O}(\beta(n))$ *and* $y = y_n + \mathcal{O}(\gamma(n))$, *with* $b\beta(n) > \gamma(n)$ *for all* n *sufficiently large. Then*

$$
\begin{aligned}
x + y &= x_n + y_n + \mathcal{O}(\beta(n) + \gamma(n)), \\
x + y &= x_n + y_n + \mathcal{O}(\beta(n)), \\
Ax &= Ax_n + \mathcal{O}(\beta(n)).
\end{aligned}
$$

In the last equation, A is an arbitrary constant, independent of n.

The proof parallels the one in the text almost perfectly, and so is omitted.

10. Use the definition of $\mathcal{O}$ to show that if $y = y_h + \mathcal{O}(h^p)$, then $hy = hy_h + \mathcal{O}(h^{p+1})$.

11. Show that if $a_n = \mathcal{O}(n^p)$ and $b_n = \mathcal{O}(n^q)$, then $a_n b_n = \mathcal{O}(n^{p+q})$.

Solution: We have

$$|a_n| \leq C_a |n^p|$$

and

$$|b_n| \leq C_b |n^q|.$$

These follow from the definition of the $\mathcal{O}$ notation. Therefore,

$$|a_n b_n| \leq C_a |n^p| |b_n| \leq (C_a |n^p|)(C_b |n^q|) = (C_a C_b)|n^{p+q}|$$

which implies that $a_n b_n = \mathcal{O}(n^{p+q})$.

12. Suppose that $y = y_h + \mathcal{O}(\beta(h))$ and $z = z_h + \mathcal{O}(\beta(h))$, for h sufficiently small. Does it follow that $y - z = y_h - z_h$ (for h sufficiently small)?

13. Show that

$$f''(x) = \frac{f(x+h) - 2f(x) + f(x-h)}{h^2} + \mathcal{O}(h^2)$$

for all h sufficiently small. Hint: Expand $f(x \pm h)$ out to the fourth order terms.

Solution: This is a straight-forward manipulation with the Taylor expansions

$$f(x+h) = f(x) + hf'(x) + \frac{1}{2}h^2 f''(x) + \frac{1}{6}h^3 f'''(x) + \frac{1}{24}h^4 f''''(\xi_1)$$

and

$$f(x-h) = f(x) - hf'(x) + \frac{1}{2}h^2 f''(x) - \frac{1}{6}h^3 f'''(x) + \frac{1}{24}h^4 f''''(\xi_2).$$

Add the two expansions to get

$$f(x+h) + f(x-h) = 2f(x) + h^2 f''(x) + \frac{1}{24}h^4 (f''''(\xi_1) + f''''(\xi_2)).$$

Now solve for $f''(x)$.

14. Explain, in your own words, why it is necessary that the constant C in (1.8) be independent of h.

1.3 A PRIMER ON COMPUTER ARITHMETIC

Exercises:

1. In each problem below, A is the exact value, and A_h is an approximation to A. Find the absolute error and the relative error.

   (a) $A = \pi$, $A_h = 22/7$;
   (b) $A = e$, $A_h = 2.71828$;
   (c) $A = \frac{1}{6}$, $A_h = 0.1667$;
   (d) $A = \frac{1}{6}$, $A_h = 0.1666$.

   Solution:

   (a) Abs. error $\leq 1.265 \times 10^{-3}$, rel. error $\leq 4.025 \times 10^{-4}$;
   (b) Abs. error $\leq 1.828 \times 10^{-6}$, rel. error $\leq 6.72 \times 10^{-7}$;
   (c) Abs. error $\leq 3.334 \times 10^{-5}$, rel. error $\leq 2.000 \times 10^{-4}$;
   (d) Abs. error $\leq 6.667 \times 10^{-5}$, rel. error $\leq 4 \times 10^{-4}$.

2. Perform the indicated computations in each of three ways: (i) Exactly; (ii) Using three-digit decimal arithmetic, with chopping; (iii) Using three-digit decimal arithmetic, with rounding. For both approximations, compute the absolute error and the relative error.

   (a) $\frac{1}{6} + \frac{1}{10}$;

   (b) $\frac{1}{6} \times \frac{1}{10}$;

   (c) $\frac{1}{9} + \left(\frac{1}{7} + \frac{1}{6}\right)$;

   (d) $\left(\frac{1}{7} + \frac{1}{6}\right) + \frac{1}{9}$.

3. For each function below explain why a naive construction will be susceptible to significant rounding error (for x near certain values), and explain how to avoid this error.

   (a) $f(x) = (\sqrt{x+9} - 3)x^{-1}$;

   (b) $f(x) = x^{-1}(1 - \cos x)$;

   (c) $f(x) = (1-x)^{-1}(\ln x - \sin \pi x)$;

   (d) $f(x) = (\cos(\pi + x) - \cos \pi)x^{-1}$;

   (e) $f(x) = (e^{1+x} - e^{1-x})(2x)^{-1}$.

   Solution: In each case, the function is susceptible to subtractive cancellation which will be amplified by division by a small number. The way to avoid the problem is to use a Taylor expansion to make the subtraction and division both explicit operations. For instance, in (a), we would write

   $$f(x) = ((3+(1/6)x - (1/216)x^2 + \mathcal{O}(x^3)) - 3)x^{-1} = (1/6) - (1/216)x + \mathcal{O}(x^2).$$

   To get greater accuracy, take more terms in the Taylor expansion.

4. For $f(x) = (e^x - 1)/x$, how many terms in a Taylor expansion are needed to get single precision accuracy (7 decimal digits) for all $x \in [0, \frac{1}{2}]$? How many terms are needed for double precision accuracy (14 decimal digits) over this same range?

5. Using single precision arithmetic, only, carry out each of the following computations, using first the form on the left side of the equals sign, then using the form on the right side, and compare the two results. Comment on what you get in light of the material in § 1.3.

   (a) $(x + \epsilon)^3 - 1 = x^3 + 3x^2\epsilon + 3x\epsilon^2 + \epsilon^3 - 1$, $x = 1.0$, $\epsilon = 0.000001$.

   (b) $-b + \sqrt{b^2 - 2c} = 2c(-b - \sqrt{b^2 - 2c})^{-1}$, $b = 1,000$, $c = \pi$.

   Solution: "Single precision" means 6 or 7 decimal digits, so the point of the problem is to do the computations using 6 or 7 digits.

   (a) Using MATLAB's `single` command on the author's laptop (running MATLAB R2019b), we get

   $$(x + \epsilon)^3 - 1 = 3.000002999797857 \times 10^{-6}$$

   but

   $$x^3 + 3x^2\epsilon + 3x\epsilon^2 + \epsilon^3 - 1 = 3.000003000019902 \times 10^{-6}.$$

(b) Using the same software and hardware, we get

$$-b + \sqrt{b^2 - 2c} = -0.003141597588410$$

but

$$2c\left(-b - \sqrt{b^2 - 2c}\right)^{-1} = -0.003141597588407.$$

What is interesting is how modern hardware and software have dramatically improved the results here. Earlier editions, which relied upon results using FORTRAN or C on a late 1990s Sun workstation, showed much more of a difference.

6. Consider the sum

$$S = \sum_{k=0}^{m} e^{-14(1-e^{-0.05k})}$$

where $m = 2 \times 10^5$. Again using only single precision, compute this two ways: First, by summing in the order indicated in the formula; second, by summing *backwards*, i.e., starting with the $k = 200,000$ term and ending with the $k = 0$ term. Compare your results and comment upon them.

7. (a) Using the computer of your choice, find three values a, b, and c, such that

$$(a + b) + c \neq a + (b + c).$$

(b) Repeat for your favorite calculator app.

(c) Do this for single precision in your preferred computing environment.

Solution: (a) The key issue is to get an approximation to the machine epsilon, then take $a = 1$, $b = c = (2/3)\mathbf{u}$ or something similar. This will guarantee that $(a + b) + c = a$ but $a + (b + c) > a$. There is an additional issue, in that MATLAB always rounds unformatted output, so to see that you got a different result you have to use (ugh!) fprintf to print out enough digits. On my laptop, I was able to use

$$a = 1$$
$$b = 1.101642356786233 \times 10^{-16}$$
$$c = 1.101642356786233 \times 10^{-16}$$

and then fprintf told me that

$$D = (a + b) + c = 1.00000000000000000,$$
$$E = a + (b + c) = 1.00000000000000022.$$

It is an interesting aspect of the history of this book, that when this exercise was first written ("a long time ago, in a computational environment far, far, away"), actual physical calculators were still commonplace (as opposed to smartphone/tablet apps). On an elderly Sharp calculator, circa 1997, the author found that $a = 1$, $b = 4 \times 10^{-10}$, and $c = 4 \times 10^{-10}$ worked. Using a scientific calculator app on his phone, the author found that $a = 1$, $b = 4 \times 10^{-16}$, and $c = 4 \times 10^{-16}$ worked.

(However, he was not able to get this to work on the Windows 10 calculator app. It would make an interesting quasi-research question to explain why.)

(b) Using MATLAB's `single` command (carefully), I used

$$a = 1$$
$$b = 3.9572964 \times 10^{-8}$$
$$c = b.$$

Then,

$$D = (a + b) + c = 1$$
$$E = a + (b + c) = 1.0000001.$$

8. Assume we are using 3-digit decimal arithmetic. For $\epsilon = 0.0001$, $a_1 = 5$, compute

$$a_2 = a_0 + \left(\frac{1}{\epsilon}\right) a_1$$

for a_0 equal to each of 1, 2, and 3. Comment.

9. Let $\epsilon \leq \mathbf{u}$. Explain, in your own words, why the computation

$$a_2 = a_0 + \left(\frac{1}{\epsilon}\right) a_1$$

is potentially rife with rounding error. (Assume that a_0 and a_1 are of comparable size.) Hint: See previous problem.

Solution: This is just a generalization of the previous problem. If ϵ is small enough, then a_2 will be independent of a_0.

10. Using the computer and language of your choice, write a program to estimate the machine epsilon.

Solution: There are lots of ways to do this. The basic idea is to add a small number to 1, and then check to see if the result is different from one, otherwise continue on. One possible solution is the following:

Algorithm 1.1

Computation of the machine epsilon.

```
x = 1.e-10;
for k=1:6000
    y = 1 + x;
    if y <= 1
    disp('macheps = ')
    disp(x)
    break
    end
    x = x*.99;
end
x
```

This produces (on the author's laptop) $\mathbf{u} = 1.101642356786233 \times 10^{-16}$. If we change the initial x to 0.5, and decrement by a factor of 2 each step, we get $\mathbf{u} = 1.110223024625157 \times 10^{-16}$, which, being larger, is a better estimate. (Why?)

11. We can compute e^{-x} using Taylor polynomials in two ways, either using

$$e^{-x} \approx 1 - x + \frac{1}{2}x^2 - \frac{1}{6}x^3 + \cdots$$

or using

$$e^{-x} \approx \frac{1}{1 + x + \frac{1}{2}x^2 + \frac{1}{6}x^3 + \cdots}.$$

Discuss, in your own words, which approach is more accurate. In particular, which one is more (or less) susceptible to rounding error?

Solution: Because of the alternating signs in the first approach, there is some concern about subtractive cancellation when it is used.

12. What is the machine epsilon for a computer that uses binary arithmetic, 24 bits for the fraction, and rounds? What if it chops?

Solution: Recall that the machine epsilon is the *largest* number x such that the computer returns $1 + x = x$. We therefore need to find the largest number x that can be represented with 24 binary digits such that $1 + x$, when rounded to 24 bits, is still equal to 1. This is perhaps best done by explicitly writing out the addition in binary notation. We have

$$1 + x = 1.000\ 0000\ 0000\ 0000\ 0000\ 0000_2$$
$$+\ 0.000\ 0000\ 0000\ 0000\ 0000\ 0000\ dddd\ dddd\ dddd\ dddd\ dddd\ dddd_2.$$

If the machine chops, then we can set all of the d values to 1 and the computer will still return $1 + x = 1$; if the machine rounds, then we need to make the first digit a zero. Thus, the desired values are

$$\mathbf{u}_{\text{round}} = \sum_{k=1}^{23} 2^{-k-24} = 0.596 \times 10^{-7},$$

and

$$\mathbf{u}_{\text{chop}} = \sum_{k=1}^{24} 2^{-k-23} = 0.119 \times 10^{-6}.$$

13. What is the machine epsilon for a computer that uses *octal* (base 8) arithmetic, assuming it retains 8 octol digits in the fraction?

◁ • • • ▷

1.4   A WORD ON COMPUTER LANGUAGES AND SOFTWARE

(No exercises in this section.)

1.5 A BRIEF HISTORY OF SCIENTIFIC COMPUTING

(No exercises in this section.)

CHAPTER 2

A SURVEY OF SIMPLE METHODS AND TOOLS

2.1 HORNER'S RULE AND NESTED MULTIPLICATION

Exercises:

1. Write each of the following polynomials in nested form.

   (a) $x^3 + 3x + 2$;
   (b) $x^6 + 2x^4 + 4x^2 + 1$;
   (c) $5x^6 + x^5 + 3x^4 + 3x^3 + x^2 + 1$;
   (d) $x^2 + 5x + 6$.

   Solution:

   (a) $x^3 + 3x + 2 = 2 + x(3 + x^2)$;
   (b) $x^6 + 2x^4 + 4x^2 + 1 = 1 + x^2(4 + x^2(2 + x^2))$;
   (c) $5x^6 + x^5 + 3x^4 + 3x^3 + x^2 + 1 = 1 + x^2(1 + x(3 + x(3 + x(1 + 5x))))$;
   (d) $x^2 + 5x + 6 = 6 + x(5 + x)$.

2. Write each of the following polynomials in nested form, but this time take advantage of the fact that they involve only even powers of x to minimize the computations.

   (a) $1 + x^2 + \frac{1}{2}x^4 + \frac{1}{6}x^6$;

(b) $1 - \frac{1}{2}x^2 + \frac{1}{24}x^4$.

3. Write each of the following polynomials in nested form.

(a) $1 - x + x^2 - x^3$;

(b) $1 - x^2 + \frac{1}{2}x^4 - \frac{1}{6}x^6$;

(c) $1 - x + \frac{1}{2}x^2 - \frac{1}{3}x^3 - \frac{1}{4}x^5$.

Solution:

(a) $1 - x + x^2 - x^3 = 1 + x((-1) + x(1 - x))$;

(b) $1 - x^2 + \frac{1}{2}x^4 - \frac{1}{6}x^6 = 1 + x^2((-1) + x^2((1/2) - (1/6)x^2))$;

(c) $1 - x + \frac{1}{2}x^2 - \frac{1}{3}x^3 - \frac{1}{4}x^5 = 1 + x((-1) + x((1/2) + x((-1/3) + x(-1/4))))$.

4. Write a computer code that takes a polynomial, defined by its coefficients, and evaluates that polynomial and its first derivative using Horner's rule. Test this code by applying it to each of the polynomials in Problem 1.

Solution: The following is a MATLAB script which does the assigned task, using the less efficient approach to computing the derivative.

```
function [y, yp] = horner1(a,x)
    n = length(a);
    y = a(n);
    for k=(n-1):(-1):1
        y = a(k) + y*x;
    end
%
    yp = (n-1)*a(n);
    for k=(n-1):(-1):2
        yp = (k-1)*a(k) + yp*x;
    end
```

5. Repeat the above, using the polynomials in Problem 2 as the test set.

Solution: The same script can be used, of course.

6. Repeat the above, using the polynomials in Problem 3 as the test set.

7. Consider the polynomial

$$p(x) = 1 + (x - 1) + \frac{1}{6}(x - 1)(x - 2) + \frac{1}{7}(x - 1)(x - 2)(x - 4).$$

This can be written in "nested-like" form by factoring out each binomial term as far as it will go, thus:

$$p(x) = 1 + (x - 1)\left(1 + (x - 2)\left(\frac{1}{6} + \frac{1}{7}(x - 4)\right)\right).$$

Write each of the following polynomials in this kind of nested form.

(a) $p(x) = 1 + \frac{1}{3}x - \frac{1}{60}x(x - 3)$;

(b) $p(x) = -1 + \frac{6}{7}(x - 1/2) - \frac{5}{21}(x - 1/2)(x - 4) + \frac{1}{7}(x - 1/2)(x - 4)(x - 2)$;

(c) $p(x) = 3 + \frac{1}{5}(x - 8) - \frac{1}{60}(x - 8)(x - 3)$.

Solution:

(a) $p(x) = 1 + \frac{1}{3}x - \frac{1}{60}x(x - 3) = 1 + x((1/3) - (1/60)(x - 3))$;

(b)

$$p(x) = -1 + \frac{6}{7}(x - 1/2) - \frac{5}{21}(x - 1/2)(x - 4) + \frac{1}{7}(x - 1/2)(x - 4)(x - 2)$$
$$= -1 + (x - (1/2))((6/7) + (x - 4)((-5/21) + (1/7)(x - 2)));$$

(c) $p(x) = 3 + \frac{1}{5}(x-8) - \frac{1}{60}(x-8)(x-3) = 3 + (x-8)((1/5) - (1/60)(x-3))$.

8. Write a computer code that computes polynomial values using the kind of nested form used in the previous problem, and test it on each of the polynomials in that problem.

9. Write a computer code to do Horner's rule on a polynomial defined by its coefficients. Test it out by using the polynomials in the previous problems. Verify that the same values are obtained when Horner's rule is used as when a naive evaluation is done.

10. Write out the Taylor polynomial of degree 5 for approximating the exponential function, using $x_0 = 0$, using the Horner form. Repeat for the degree 5 Taylor approximation to the sine function. (Be sure to take advantage of the fact that the Taylor expansion to the sine uses only odd powers.)

    Solution: For the exponential function, we get

    $$p_5(x) = 1 + x(1 + x((1/2) + x((1/6) + x((1/24) + (1/120)x))));$$

    for the sine function we get

    $$p_5(x) = x(1 + x^2((-1/6) + (1/120)x^2)).$$

11. For each function in Problem 11 of §1.1, write the polynomial approximation in Horner form, and use this as the basis for a computer program that approximates the function. Compare the accuracy you actually achieve (based on the built-in intrinsic functions on your computer) to that which was theoretically established. Be sure to check that the required accuracy is achieved over the entire interval in question.

12. Repeat the above, except this time compare the accuracy of the *derivative* approximation constructed by taking the derivative of the approximating polynomial. Be sure to use the derivative form of Horner's rule to evaluate the polynomial.

◁ • • • ▷

2.2   DIFFERENCE APPROXIMATIONS TO THE DERIVATIVE

Exercises:

1. Use the methods of this section to show that

$$f'(x) = \frac{f(x) - f(x - h)}{h} + \mathcal{O}(h).$$

Solution: We have, for any a, that

$$f(x + a) = f(x) + af'(x) + \frac{1}{2}a^2 f''(\xi).$$

Therefore, taking $a = -h$,

$$f(x + a) \quad = \quad f(x - h) = f(x) - hf'(x) + \frac{1}{2}h^2 f''(\xi)$$

$$\Rightarrow \quad f'(x) = \frac{f(x) - f(x - h)}{h} + \frac{1}{2}hf''(\xi).$$

2. Compute, by hand, approximations to $f'(1)$ for each of the following functions, using $h = 1/16$ and each of the derivative approximations contained in (2.1) and (2.5).

   (a) $f(x) = \sqrt{x + 1}$;
   (b) $f(x) = \arctan x$;
   (c) $f(x) = \sin \pi x$;
   (d) $f(x) = e^{-x}$;
   (e) $f(x) = \ln x$.

3. Write a computer program which uses the same derivative approximations as in the previous problem to approximate the first derivative at $x = 1$ for each of the following functions, using $h^{-1} = 4, 8, 16, 32$. Verify that the predicted theoretical accuracy is obtained.

   (a) $f(x) = \sqrt{x + 1}$;
   (b) $f(x) = \arctan x$;
   (c) $f(x) = \sin \pi x$;
   (d) $f(x) = e^{-x}$;
   (e) $f(x) = \ln x$.

   Solution: I wrote a simple FORTRAN program to do this for the single case of (b). The results I got, using double precision, are in Table 2.1. Note that the error goes down by a factor of 2 for $D_1(h)$, and a factor of 4 for $D_2(h)$, thus confirming the theoretical accuracy.

4. Use the approximations from this section to fill in approximations to the missing values in Table 2.2 (Table 2.4 in the text).

Table 2.1   Derivative Approximations.

h^{-1}	$D_1(h)$	Error	Ratio	$D_2(h)$	Error	Ratio
4	0.44262888	0.573711×10^{-1}	0.000	0.50510855	0.510855×10^{-2}	0.000
8	0.47004658	0.299534×10^{-1}	1.915	0.50129595	0.129595×10^{-2}	3.942
16	0.48470016	0.152998×10^{-1}	1.958	0.50032514	0.325139×10^{-3}	3.986
32	0.49226886	0.773114×10^{-2}	1.979	0.50008136	0.813564×10^{-4}	3.996
64	0.49611409	0.388591×10^{-2}	1.990	0.50002034	0.203436×10^{-4}	3.999
128	0.49805196	0.194804×10^{-2}	1.995	0.50000509	0.508617×10^{-5}	4.000
256	0.49902471	0.975291×10^{-3}	1.997	0.50000127	0.127156×10^{-5}	4.000
512	0.49951204	0.487963×10^{-3}	1.999	0.50000032	0.317891×10^{-6}	4.000
1024	0.49975594	0.244061×10^{-3}	1.999	0.50000008	0.794728×10^{-7}	4.000

Table 2.2   Table for Problem 4.

x	$f(x)$	$f'(x)$
1.00	1.0000000000	
1.10	0.9513507699	
1.20	0.9181687424	
1.30	0.8974706963	
1.40	0.8872638175	
1.50	0.8862269255	
1.60	0.8935153493	
1.70	0.9086387329	
1.80	0.9313837710	
1.90	0.9617658319	
2.00	1.0000000000	

5. Use the error estimate (2.5) for the centered difference approximation to the first derivative to prove that this approximation will be *exact* for any quadratic polynomial.

   Solution: We have that

   $$f'(x) - \frac{f(x+h) - f(x-h)}{2h} = Ch^2 f'''(\xi),$$

   for some ξ between $x - h$ and $x + h$. If f is a quadratic, then it has the general form

   $$f(x) = ax^2 + bx + c,$$

   for given constants a, b, and c. But the third derivative of this kind of function will always be identically zero.

6. Find coefficients A, B, and C so that

   (a) $f'(x) = Af(x) + Bf(x+h) + Cf(x+2h) + \mathcal{O}(h^2)$;

   (b) $f'(x) = Af(x) + Bf(x-h) + Cf(x-2h) + \mathcal{O}(h^2)$.

   Hint: Use Taylor's theorem.

Solution: Write out Taylor expansions for $f(x + h)$ and $f(x + 2h)$, thus:

$$f(x + h) = f(x) + hf'(x) + (h^2/2)f''(x) + \mathcal{O}(h^3)$$

and

$$f(x + 2h) = f(x) + 2hf'(x) + 2h^2 f''(x) + \mathcal{O}(h^3).$$

Multiply the first one by 4, and subtract the two of them, to get

$$4f(x + h) - f(x + 2h) = 3f(x) + 2hf'(x) + \mathcal{O}(h^3);$$

solve this for f' to get

$$f'(x) = (-3/2h)f(x) + (2/h)f(x + h) + (-1/2h)f(x + 2h) + \mathcal{O}(h^2)$$

so that $A = -3/2h$, $B = 2/h$, and $C = -1/2h$. A similar approach works for (b), with $A = 3/2$, $B = -2/h$, and $C = 1/h$.

7. Fill in the data from Problem 4 using methods that are $\mathcal{O}(h^2)$ at each point. Hint: See the previous problem.

8. Use Taylor's Theorem to show that the approximation

$$f'(x) \approx \frac{8f(x + h) - 8f(x - h) - f(x + 2h) + f(x - 2h)}{12h}$$

is $\mathcal{O}(h^4)$.

Solution: We have

$$f(x+h) = f(x)+hf'(x)+(h^2/2)f''(x)+(h^3/6)f'''(x)+(h^4/24)f''''(x)+\mathcal{O}(h^5),$$

$$f(x+2h) = f(x)+2hf'(x)+2h^2 f''(x)+(4h^3/3)f'''(x)+(2h^4/3)f''''(x)+\mathcal{O}(h^5),$$

$$f(x-h) = f(x)-hf'(x)+(h^2/2)f''(x)-(h^3/6)f'''(x)+(h^4/24)f''''(x)+\mathcal{O}(h^5),$$

and

$$f(x-2h) = f(x)-2hf'(x)+2h^2 f''(x)-(4h^3/3)f'''(x)+(2h^4/3)f''''(x)+\mathcal{O}(h^5).$$

Therefore,

$$8f(x + h) - 8f(x - h) = 16hf'(x) + (8h^3/3)f'''(x) + \mathcal{O}(h^5)$$

and

$$f(x + 2h) - f(x - 2h) = 4hf'(x) + (8h^3/3)f'''(x) + \mathcal{O}(h^5).$$

Hence,

$$8f(x + h) - 8f(x - h) - f(x + 2h) + f(x - 2h) = 12hf'(x) + \mathcal{O}(h^5),$$

so that, solving for f', we get the desired result.

9. Use the derivative approximation from Problem 8 to approximate $f'(1)$ for the same functions as in Problem 3. Verify that the expected rate of decrease is observed for the error.

10. Use the derivative approximation from Problem 8 to fill in as much as possible of the table in Problem 4.

    Solution: Because this formula uses more points that are farther from the point of interest than did the previous formulas, we cannot do as much of the table as before. We get:

$$
\begin{aligned}
f'(1.2) &\approx -0.2652536725, \\
f'(1.3) &\approx -0.1517629626, \\
f'(1.4) &\approx -0.05441397741, \\
f'(1.5) &\approx 0.03237018117, \\
f'(1.6) &\approx 0.1126454212, \\
f'(1.7) &\approx 0.1895070563, \\
f'(1.8) &\approx 0.2654434511.
\end{aligned}
$$

11. Let $f(x) = \arctan x$. Use the derivative approximation from Problem 8 to approximate $f'(\frac{1}{4}\pi)$ using $h^{-1} = 2, 4, 8, \ldots$. Try to take h small enough that the rounding error effect begins to dominate the mathematical error. For what value of h does this begin to occur? (You may have to restrict yourself to working in single precision.)

12. Use Taylor expansions for $f(x \pm h)$ to derive an $\mathcal{O}(h^2)$ accurate approximation to f'' using $f(x)$ and $f(x \pm h)$. Provide all the details of the error estimate. Hint: Go out as far as the fourth derivative term, and then add the two expansions.

    Solution: We have

$$
f(x + h) = f(x) + hf'(x) + (h^2/2)f''(x) + (h^3/6)f'''(x) + \mathcal{O}(h^4),
$$

and

$$
f(x - h) = f(x) - hf'(x) + (h^2/2)f''(x) - (h^3/6)f'''(x) + \mathcal{O}(h^4),
$$

so that

$$
f(x + h) + f(x - h) = 2f(x) + h^2 f''(x) + \mathcal{O}(h^4).
$$

Thus, solving for the second derivative yields

$$
f''(x) = \frac{f(x + h) - 2f(x) + f(x - h)}{h^2} + \mathcal{O}(h^2).
$$

13. Let $h > 0$ and $\eta > 0$ be given, where $\eta = \theta h$, for $0 < \theta < 1$. Let f be some smooth function. Use Taylor expansions for $f(x + h)$ and $f(x - \eta)$ in terms of f and its derivatives at x in order to construct an approximation to $f'(x)$ that depends on $f(x + h)$, $f(x)$, and $f(x - \eta)$, and which is $\mathcal{O}(h^2)$ accurate. Check your work by verifying that for $\theta = 1 \Rightarrow \eta = h$ you get the same results as in the text.

    Solution: We write the two Taylor expansions

$$
f(x + h) = f(x) + hf'(x) + \frac{1}{2}h^2 f''(x) + \frac{1}{6}h^3 f'''(c_1)
$$

and

$$f(x - \eta) = f(x) - \eta f'(x) + \frac{1}{2}\eta^2 f''(x) - \frac{1}{6}\eta^3 f'''(c_2).$$

If we simply subtract these, the second derivative terms will not cancel out (because $h \neq \eta$), so we have to multiply the first expansion by η^2 and the second one by h^2 to get

$$\eta^2 f(x + h) = \eta^2 f(x) + \eta^2 h f'(x) + \frac{1}{2}\eta^2 h^2 f''(x) + \frac{1}{6}\eta^2 h^3 f'''(c_1)$$

and

$$h^2 f(x - \eta) = h^2 f(x) - h^2 \eta f'(x) + \frac{1}{2}h^2 \eta^2 f''(x) - \frac{1}{6}h^2 \eta^3 f'''(c_2).$$

Now subtract to get

$$\eta^2 f(x + h) - h^2 f(x - \eta) = (\eta^2 - h^2) f(x) + (\eta^2 h - h^2 \eta) f'(x) + \frac{1}{6}(\eta^2 h^3 f'''(c_1)$$
$$+ h^2 \eta^3 f'''(c_2)).$$

Solve for $f'(x)$ to get

$$f'(x) = \frac{\eta^2 f(x + h) - (\eta^2 - h^2) f(x) - h^2 f(x - \eta)}{\eta^2 h - h^2 \eta} - \frac{1}{6}\frac{\eta^2 h^3 f'''(c_1) + h^2 \eta^3 f'''(c_2)}{\eta^2 h - h^2 \eta}.$$

Because we assumed $\eta = \theta h$, there is substantial simplification that we can do to get

$$f'(x) = \frac{\theta^2 f(x + h) - (\theta^2 - 1) f(x) - f(x - \eta)}{\theta^2 h - h\theta} - \frac{1}{6}\frac{\theta^2 h^2 f'''(c_1) + h^2 \theta^3 f'''(c_2)}{\theta^2 - \theta},$$

which is sufficient to establish the $\mathcal{O}(h^2)$ estimate for the error.

14. Write a computer program to test the approximation to the second derivative from Problem 12 by applying it to estimate $f''(1)$ for each of the following functions, using $h^{-1} = 4, 8, 16, 32$. Verify that the predicted theoretical accuracy is obtained.

    (a) $f(x) = e^{-x}$;

    (b) $f(x) = \cos \pi x$;

    (c) $f(x) = \sqrt{1 + x}$;

    (d) $f(x) = \ln x$.

15. Define the following function:

$$f(x) = \ln\left(e^{\sqrt{x^2 + 1}} \sin(\pi x) + \tan \pi x\right).$$

Compute values of f' over the range $[0, \frac{1}{4}]$ using two methods:

    (a) Using the centered-difference formula from (2.5);

    (b) Using ordinary calculus to find the formula for f'.

Comment.

Solution: The function f is designed to make direct computation of f' a difficult task for the average undergraduate student, thus highlighting that there are times when being able to approximate the derivative as was done in this section might be preferable to doing the exact computation.

16. Let $f(x) = e^x$, and consider the problem of approximating $f'(1)$, as in the text. Let $D_1(h)$ be the difference approximation in (2.1). Using the appropriate values in Table 2.1, compute the new approximations

$$\Delta_1(h) = 2D_1(h) - D_1(2h);$$

and compare these values to the exact derivative value. Are they more or less accurate that the corresponding values of D_1? Try to deduce, from your calculations, how the error depends on h.

Solution: For the sake of completeness we have reproduced the table from the text. Using those values we can compute

$$\Delta_1(0.25) = 2D_1(0.25) - D_1(.5) = 2.649674415 \Rightarrow e^1 - \Delta_1(0.25) = 0.068607413;$$
$$\Delta_1(0.125) = 2D_1(0.125) - D_1(0.25) = 2.702717782 \Rightarrow e^1 - \Delta_1(0.125) = 0.015564046;$$
$$\Delta_1(0.0625) = 2D_1(0.0625) - D_1(0.125) = 2.714572906 \Rightarrow e^1 - \Delta_1(0.0625) = 0.003708922;$$
$$\Delta_1(0.03125) = 2D_1(0.03125) - D_1(0.0625) = 2.717372894 \Rightarrow e^1 - \Delta_1(0.03125) = 0.00090834.$$

The error appears to be going down by a factor of 4 as h is cut in half.

Table 2.3   (Table 2.1 in text.) Example of derivative approximation to $f(x) = e^x$ at $x = 1$.

h^{-1}	$D_1(h)$	$E_1(h) = f'(1) - D_1(h)$	$D_2(h)$	$E_2(h) = f'(1) - D_2(h)$
2	3.526814461	-0.8085327148	2.832967758	-0.1146860123
4	3.088244438	-0.3699626923	2.746685505	$-0.2840375900 \times 10^{-1}$
8	2.895481110	-0.1771993637	2.725366592	$-0.7084846497 \times 10^{-2}$
16	2.805027008	$-0.8674526215 \times 10^{-1}$	2.720052719	$-0.1770973206 \times 10^{-2}$
32	2.761199951	$-0.4291820526 \times 10^{-1}$	2.718723297	$-0.4415512085 \times 10^{-3}$
64	2.739639282	$-0.2135753632 \times 10^{-1}$	2.718391418	$-0.1096725464 \times 10^{-3}$
128	2.728942871	$-0.1066112518 \times 10^{-1}$	2.718307495	$-0.2574920654 \times 10^{-4}$

17. Repeat the above idea for $f(x) = \arctan(x)$, $x = 1$ (but this time you will have to compute the original $D_1(h)$ values).

18. By keeping more terms in the Taylor expansion for $f(x + h)$, show that the error in the derivative approximation (2.1) can be written as

$$f'(x) - \left(\frac{f(x + h) - f(x)}{h} \right) = -\frac{1}{2}hf''(x) - \frac{1}{6}h^2 f'''(x) - \cdots \qquad (2.1)$$

Use this to construct a derivative approximation involving $f(x)$, $f(x + h)$, and $f(x + 2h)$ that is $\mathcal{O}(h^2)$ accurate. Hint: Use (2.6) to write down the error in the approximation

$$f'(x) \approx \frac{f(x + 2h) - f(x)}{2h}$$

and combine the two error expansions so that the terms that are $\mathcal{O}(h)$ are eliminated.

Solution: We have

$$f(x+h) = f(x) + hf'(x) + (1/2)h^2 f''(x) + \mathcal{O}(h^3)$$

so that we can get

$$f'(x) = \frac{f(x+h) - f(x)}{h} - (1/2)hf''(x) + \mathcal{O}(h^2),$$

or, equivalently,

$$D_1(h) = \frac{f(x+h) - f(x)}{h} = f'(x) + (1/2)hf''(x) + \mathcal{O}(h^2).$$

Therefore,

$$D_1(2h) = \frac{f(x+2h) - f(x)}{2h} = f'(x) + (1/2)(2h)f''(x) + \mathcal{O}(h^2).$$

Thus,

$$2D_1(2h) - D_1(h) = 2\left(f'(x) + (1/2)(2h)f''(x) + \mathcal{O}(h^2)\right)$$
$$- \left(f'(x) + (1/2)hf''(x) + \mathcal{O}(h^2)\right)$$

so that

$$2D_1(2h) - D_1(h) = f'(x) + \mathcal{O}(h^2).$$

19. Apply the method derived above to the list of functions in Problem 3, and confirm that the method is as accurate in practice as is claimed.

20. Let $f(x) = e^x$, and consider the problem of approximating $f'(1)$, as in the text. Let $D_2(h)$ be the difference approximation in (2.5). Using the appropriate values in Table 2.3, compute the new approximations

$$\Delta_2(h) = (4D_2(h) - D_2(2h))/3;$$

and compare these values to the exact derivative value. Are they more or less accurate that the corresponding values of D_2? Try to deduce, from your calculations, how the error depends on h.

Solution: The $D_2(h)$ values are

$$D_2(1/2) = 2.832967758$$

$$D_2(1/4) = 2.746685505$$

$$D_2(1/8) = 2.725366592$$

$$D_2(1/16) = 2.720052719$$

$$D_2(1/32) = 2.718723297$$

$$D_2(1/64) = 2.718391418$$

$$D_2(1/128) = 2.718307495$$

from which we get that

$$\Delta_2(1/4) = 2.717924754$$

$$\Delta_2(1/8) = 2.718260288$$

$$\Delta_2(1/16) = 2.718281428$$

$$\Delta_2(1/32) = 2.718280156$$

$$\Delta_2(1/64) = 2.718280792$$

$$\Delta_2(1/128) = 2.718279521.$$

These values are, initially, much more accurate than $D_2(h)$, although rounding error begins to corrupt the computation. It can be shown that this approximation is $\mathcal{O}(h^4)$.

21. Repeat the above idea for $f(x) = \arctan(x)$, $x = 1$ (but this time you will have to compute the original $D_2(h)$ values).

22. The same ideas as in Problem 18 can be applied to the centered difference approximation (2.5). Show that in this case the error satisfies

$$f'(x) - \frac{f(x+h) - f(x-h)}{2h} = -\frac{1}{6}h^2 f'''(x) - \frac{1}{120}h^4 f'''''(x) - \cdots \quad (2.2)$$

Use this to construct a derivative approximation involving $f(x \pm h)$, and $f(x \pm 2h)$ that is $\mathcal{O}(h^4)$ accurate.

Solution: Since

$$f(x+h) = f(x) + hf'(x) + (1/2)h^2 f''(x) + (1/6)h^3 f'''(x) + (1/24)h^4 f''''(x) + \mathcal{O}(h^5)$$

and

$$f(x-h) = f(x) - hf'(x) + (1/2)h^2 f''(x) - (1/6)h^3 f'''(x) + (1/24)h^4 f''''(x) + \mathcal{O}(h^5),$$

we have that

$$f(x+h) - f(x-h) = 2hf'(x) + (1/3)h^3 f'''(x) + \mathcal{O}(h^5)$$

or,

$$f'(x) - \frac{f(x+h) - f(x-h)}{2h} = (1/6)h^2 f'''(x) + \mathcal{O}(h^4).$$

Therefore,

$$f'(x) - \frac{f(x+2h) - f(x-2h)}{4h} = (4/6)h^2 f'''(x) + \mathcal{O}(h^4),$$

so that

$$4\left(f'(x) - \frac{f(x+h) - f(x-h)}{2h}\right) - \left(f'(x) - \frac{f(x+2h) - f(x-2h)}{4h}\right) = \mathcal{O}(h^4).$$

We can manipulate with the left side to get that

$$3f'(x) - \frac{8(f(x+h) - f(x-h)) - (f(x+2h) - f(x-2h))}{4h} = \mathcal{O}(h^4),$$

so that finally we have

$$f'(x) - \frac{8(f(x+h) - f(x-h)) - (f(x+2h) - f(x-2h))}{12h} = \mathcal{O}(h^4).$$

This is simply an alternate derivation of the method from Exercise 8.

23. Apply the method derived above to the list of functions in Problem 3, and confirm that the method is as accurate in practice as is claimed.

24. What if the grid spacings are not equal? Suppose that we have

$$x_{i+1} - x_i = H$$

and

$$x_i - x_{i-1} = h.$$

Can you construct a weighted average of one-sided approximations to $f'(x_i)$ that is second-order accurate? (You will need to keep more terms in the Taylor expansion than we did in the text.)

◁ ● ● ▷

2.3 APPLICATION: EULER'S METHOD FOR INITIAL VALUE PROBLEMS

Exercises:

1. Use Euler's method with $h = 0.25$ to compute approximate solution values for the initial value problem

$$y' = \sin(t + y), \quad y(0) = 1.$$

You should get $y_4 = 1.851566895$ (be sure your calculator is set in radians).

Solution:

$$y_1 = y_0 + (0.25)\sin(0 + y_0) = 1 + 0.25(0.8414709848) = 1.210367746;$$

$$y_2 = y_1 + (0.25)\sin(h + y_1) = 1.458844986;$$

$$y_3 = y_2 + (0.25)\sin(2h + y_2) = 1.690257280;$$

$$y_4 = y_3 + (0.25)\sin(3h + y_3) = 1.851566896.$$

2. Repeat the above with $h = 0.20$. What value do you now get for $y_5 \approx y(1)$?

3. Repeat the above with $h = 0.125$. What value do you now get for $y_8 \approx y(1)$?

Solution:

$$y_1 = y_0 + (0.125)\sin(0 + y_0) = 1 + 0.125(0.8414709848) = 1.105183873;$$

$$y_2 = y_1 + (0.125)\sin(h + y_1) = 1.223002653;$$

$$y_3 = y_2 + (0.125)\sin(2h + y_2) = 1.347405404;$$

$$y_4 = y_3 + (0.125) \sin(3h + y_3) = 1.470971572;$$
$$y_5 = y_4 + (0.125) \sin(4h + y_4) = 1.586095664;$$
$$y_6 = y_5 + (0.125) \sin(5h + y_5) = 1.686335284;$$
$$y_7 = y_6 + (0.125) \sin(6h + y_6) = 1.767364012;$$
$$y_8 = y_7 + (0.125) \sin(7h + y_7) = 1.827207570.$$

4. Use Euler's method with $h = 0.25$ to compute approximate solution values for

$$y' = e^{t-y}, \quad y(0) = -1.$$

What approximate value do you get for $y(1) = 0.7353256638$?

5. Repeat the above with $h = 0.20$. What value do you now get for $y_5 \approx y(1)$?

   Solution: We have the computation

$$y_{n+1} = y_n + he^{t_n - y_n},$$

where $h = 0.2$, $t_0 = 0$, and $y_0 = -1$. Hence

$$y_1 = y_0 + he^{t_0 - y_0} = -1 + (0.2)e^{0-(-1)} = -0.4563436344;$$

$$y_2 = y_1 + he^{t_1 - y_1} = -0.4563436344 + (0.2)e^{0.2-(-0.4563436344)} = -0.0707974456;$$

$$y_3 = y_2 + he^{t_2 - y_2} = -0.0707974456 + (0.2)e^{0.4-(-0.0707974456)} = 0.2494566764;$$

$$y_4 = y_3 + he^{t_3 - y_3} = 0.2494566764 + (0.2)e^{0.6-(0.2494566764)} = 0.5334244306;$$

$$y_5 = y_4 + he^{t_4 - y_4} = 0.5334244306 + (0.2)e^{0.8-(0.5334244306)} = 0.7945216786.$$

6. Repeat the above with $h = 0.125$. What value do you now get for $y_8 \approx y(1)$?

7. Use Euler's method with $h = 0.0625$ to compute approximate solution values over the interval $0 \le t \le 1$ for the initial value problem

$$y' = t - y, \quad y(0) = 2,$$

which has exact solution $y(t) = 3e^{-t} + t - 1$. Plot your approximate solution as a function of t, and plot the error as a function of t.

   Solution: The plots are given in Figures 2.1 and 2.2.

8. Repeat the above for the equation

$$y' = e^{t-y}, \quad y(0) = -1,$$

which has exact solution $y = \ln(e^t - 1 + e^{-1})$.

9. Repeat the above for the equation

$$y' + y = \sin t, \quad y(0) = -1,$$

which has exact solution $y = (\sin t - \cos t - e^{-t})/2$.

   Solution: The plots are given in Figures 2.3 and 2.4.

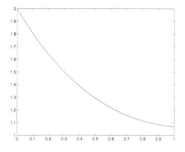

Figure 2.1   Exact solution for Exercise 2.3.7.

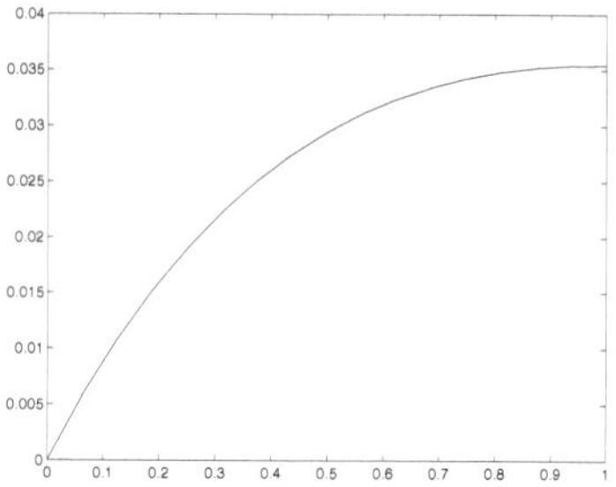

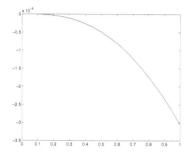

Figure 2.2   Error plot for Exercise 2.3.7.

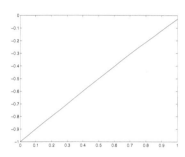

Figure 2.3   Exact solution for Exercise 2.3.9.

Figure 2.4   Error plot for Exercise 2.3.9.

10. Use Euler's method to compute approximate solutions to each of the initial value problems below, using $h^{-1} = 2, 4, 8, 16$. Compute the maximum error over the interval $[0, 1]$ for each value of h. Plot your approximate solutions for the $h = 1/16$ case. Hint: Verify that your code works by using it to reproduce the results given for the examples in the text.

(a) $y' = t - y$, $y(0) = 2$; $y(t) = 3e^{-t} + t - 1$;

(b) $y' + 4y = 1$, $y(0) = 1$; $y(t) = \frac{1}{4}(3e^{-4t} + 1)$;

(c) $y' = -y \ln y$, $y(0) = 3$; $y(t) = e^{(\ln 3)e^{-t}}$.

11. Consider the approximate values in Tables 2.5 and 2.6 in the text. Let y_k^8 denote the approximate values for $h = 1/8$, and y_k^{16} denote the approximate values for $h = 1/16$. Note that

$$y_k^8 \approx y(k/8)$$

and

$$y_{2k}^{16} \approx y(2k/16) = y(k/8) \approx y_k^8,$$

thus y_k^8 and y_{2k}^{16} are both approximations to the same value. Compute the set of new approximations

$$u_k = 2y_{2k}^{16} - y_k^8$$

and compare these to the corresponding exact solution values. Are they better or worse as an approximation?

Solution: Table 2.4 shows the new solution and the error, which is much smaller than that obtained using Euler's method directly.

Table 2.4   Solution values for Exercise 2.3.11.

t_k	u_k	$y(t_k) - u_k$
0.000	-1.00000000000000	0
0.125	-0.87500508526970	-0.00000483294465
0.250	-0.75013901623461	-0.00001560652915
0.375	-0.62573474516817	-0.00002744064045
0.500	-0.50230696317037	-0.00003687832904
0.625	-0.38052202438375	-0.00004161315812
0.750	-0.26116806849085	-0.00004026230490
0.875	-0.14512551018440	-0.00003217761850
1.000	-0.03333809297284	-0.00001728814300

12. Apply the basic idea from the previous problem to the approximation of solutions to

$$y' = e^{t-y}, \quad y(0) = -1,$$

which has exact solution $y = \ln(e^t - 1 + e^{-1})$.

13. Assume that the function f satisfies

$$|f(t, y) - f(t, z)| \le K|y - z|$$

for some constant K. Use this and (2.9)-(2.10) to show that the error $|y(t_{n+1}) - y_{n+1}|$ satisfies the recursion,

$$|y(t_{n+1}) - y_{n+1}| \le (1 + Kh)|y(t_n) - y_n| + \frac{1}{2}h^2 Y_2$$

where

$$Y_2 = \max_t |y''(t)|.$$

Solution: If we take $t = t_n$ and subtract (2.10) from (2.9) we get

$$y(t_n + h) - y_{n+1} = y(t_n) - y_n + h(f(t_n, y(t_n)) - f(t_n, y_n)),$$

from which we have

$$|y(t_{n+1}) - y_{n+1}| \le |y(t_n) - y_n| + hK|y(t_n) - y_n)|,$$

and the desired result follows immediately.

◁ • • • ▷

2.4 LINEAR INTERPOLATION

Exercises:

1. Use linear interpolation to find approximations to the following values of the error function, using the table in the text. For each case, give an upper bound on the error in the approximation:

    (a) $\mathrm{erf}(0.56)$;

    (b) $\mathrm{erf}(0.07)$;

    (c) $\mathrm{erf}(0.34)$;

    (d) $\mathrm{erf}(0.12)$;

    (e) $\mathrm{erf}(0.89)$.

    Solution: For Part (a), we have $x_0 = 0.5$ and $x_1 = 0.6$, so

    $$p_1(x) = \left(\frac{x - 0.5}{0.6 - 0.5}\right)(0.60385609084793) + \left(\frac{0.6 - x}{0.6 - 0.5}\right)(0.52049987781305),$$

    thus

    $$\begin{aligned} p_1(0.56) &= \left(\frac{0.06}{0.1}\right)(0.60385609084793) + \left(\frac{0.04}{0.1}\right)(0.52049987781305) \\ &= 0.5705136056. \end{aligned}$$

    The upper bound on the error comes straight from Theorem 2.1. We have

    $$|\mathrm{erf}(x) - p_1(x)| \le \frac{1}{8}(0.1)^2 \max|f''(t)|,$$

    where the maximum of the second derivative is computed over the interval $[0.5, 0.6]$. We have

    $$(\mathrm{erf}(x))'' = \frac{-4x}{\sqrt{\pi}} e^{-x^2}$$

    so that

    $$|(\mathrm{erf}(x))''| \le \frac{4 \times 0.6}{\sqrt{\pi}} e^{-0.25} = 1.05\ldots .$$

    Hence the error satisfies

    $$|\mathrm{erf}(0.56) - p_1(0.56)| \le \frac{0.01 \times 1.05\ldots}{8} = 0.00132\ldots .$$

2. The gamma function, denoted by $\Gamma(x)$, occurs in a number of applications, most notably probability theory and the solution of certain differential equations. It is basically the generalization of the factorial function to non-integer values, in that $\Gamma(n + 1) = n!$. Table 2.5 (Table 2.8 in the text) gives values of $\Gamma(x)$ for x between 1 and 2. Use linear interpolation to approximate values of $\Gamma(x)$ as given below:

    (a) $\Gamma(1.290) = 0.8990415863$;

    (b) $\Gamma(1.005) = 0.9971385354$;

Table 2.5   (Table 2.8 in the text) Table of $\Gamma(x)$ values.

x	$\Gamma(x)$
1.00	1.0000000000
1.10	0.9513507699
1.20	0.9181687424
1.30	0.8974706963
1.40	0.8872638175
1.50	0.8862269255
1.60	0.8935153493
1.70	0.9086387329
1.80	0.9313837710
1.90	0.9617658319
2.00	1.0000000000

(c) $\Gamma(1.930) = 0.9723969178$;

(d) $\Gamma(1.635) = 0.8979334930$.

3. Theorem 2.1 requires an upper bound on the second derivative of the function being interpolated, and this is not always available as a practical matter. However, if a table of values is available, we can use a difference approximation to the second derivative to *estimate* the upper bound on the derivative and hence the error. Recall, from Problem 12 of §2.2,

$$f''(x) \approx \frac{f(x - h) - 2f(x) + f(x + h)}{h^2}.$$

Assume the function values are given at the equally spaced grid points $x_k = a + kh$ for some grid spacing, h. Using the estimate

$$\max_{x_k \leq x \leq x_{k+1}} |f''(x)| \approx \max \left\{ \frac{f(x_{k-1}) - 2f(x_k) + f(x_{k+1})}{h^2}, \frac{f(x_k) - 2f(x_{k+1}) + f(x_{k+2})}{h^2} \right\}$$

to approximate the derivative upper bound, estimate the error made in using linear interpolation to approximate each of the following values of $\Gamma(x)$, based on the same table as in the previous problem:

(a) $\Gamma(1.290) = 0.8990415863$;

(b) $\Gamma(1.579) = 0.8913230181$;

(c) $\Gamma(1.456) = 0.8856168100$;

(d) $\Gamma(1.314) = 0.8954464400$;

(e) $\Gamma(1.713) = 0.9111663772$.

Solution: We do Part (d), only. This is a "two-stage" problem. First, we compute an approximate value for $\Gamma(1.314)$ using linear interpolation. Then we estimate the error using the suggested device for approximating the second derivative.

The interpolation is straight-forward:

$$p_1(x) = \left(\frac{x - 1.3}{1.4 - 1.3}\right)(0.8872638175) + \left(\frac{1.4 - x}{1.4 - 1.3}\right)(0.8974706963),$$

so that

$$\Gamma(1.314) \approx p_1(1.314) = \left(\frac{0.014}{0.1}\right)(0.8872638175) + \left(\frac{0.086}{0.1}\right)(0.8974706963)$$
$$= 0.8960417333.$$

Now, the error is bounded according to

$$|\Gamma(x) - p_1(x)| \leq \frac{1}{8}(x_1 - x_0)^2 \max |(\Gamma(t))''|,$$

where the maximum is taken over the interval $[x_0, x_1]$. We don't have a formula for $\Gamma(x)$, so we can't get one for the second derivative. But we do have the approximation (for any function)

$$f''(x) \approx \frac{f(x + h) - 2f(x) + f(x - h)}{h^2}$$

so we use this to estimate the second derivative of $\Gamma(x)$:

$$\Gamma''(1.3) \approx \frac{0.8872638175 - 2(0.8974706963) + 0.9181687424}{0.01} = 1.049\ldots$$

$$\Gamma''(1.4) \approx \frac{0.8862269255 - 2(0.8872638175) + 0.8974706963}{0.01} = 0.917\ldots$$

$$\max |\Gamma''(t)| \approx \max\{1.049\ldots, 0.917\ldots\} = 1.049\ldots.$$

Therefore the estimate on the error is

$$|\Gamma(1.314) - p_1(1.314)| \leq \frac{1}{8}(0.1)^2(1.049\ldots) = 0.00131\ldots.$$

According to MATLAB, the exact value (to 15 places) is

$$\Gamma(1.314) = 0.89544644002887,$$

thus the interpolation error is -5.95×10^{-4}, so our estimate is a tad high, but certainly it is within the range of acceptable estimation.

4. Construct a linear interpolating polynomial to the function $f(x) = x^{-1}$ using $x_0 = \frac{1}{2}$ and $x_1 = 1$ as the nodes. What is the upper bound on the error over the interval $[\frac{1}{2}, 1]$, according to the error estimate?

5. Repeat the above for $f(x) = \sqrt{x}$, using the interval $[\frac{1}{4}, 1]$.

   Solution: The polynomial is

$$p_1(x) = \frac{x - 1/4}{1 - 1/4}(1) + \frac{1 - x}{1 - 1/4}(1/2) = (2x + 1)/3.$$

The error bound is given by

$$|f(x) - p_1(x)| \le \frac{1}{8}(3/4)^2 \max_{t \in [1/4,1]} |(-1/4)t^{-3/2}| = (9/128) \times (1/4) \times (1/4)^{-3/2}$$

$$= 9/64 = 0.140625.$$

This is again a conservative estimate, since a simple plot of the difference $\sqrt{x} - p_1(x)$ shows that the maximum absolute error is about 0.04.

6. Repeat the above for $f(x) = x^{1/3}$, using the interval $[\frac{1}{8}, 1]$.

7. If we want to use linear interpolation to the sine function and obtain an accuracy of 10^{-6}, how close together do the entries in the table have to be? What if we change the error criterion to 10^{-3}?

   Solution: This amounts to asking how small does $x_1 - x_0$ have to be to make the upper bound in the error estimate less than the specified tolerance. For convenience set $h = x_1 - x_0$. We have, then, for $f(x) = \sin x$,

   $$|f(x) - p_1(x)| \le \frac{1}{8}h^2(1)$$

   so that making the error less than 10^{-6} requires taking $h \le \sqrt{8} \times 10^{-3} = 0.2828427124 \times 10^{-2}$. For an error less than 10^{-3} we need $h \le 0.8944271912 \times 10^{-1}$.

8. Repeat the above for $f(x) = \tan x$, for $x \in [-\pi/4, \pi/4]$.

9. If we try to approximate the logarithm by using a table of logarithm entries, together with linear interpolation, to construct the approximation to $\ln(1+x)$ over the interval $[-\frac{1}{2}, 0]$, how many points are required to get an approximation that is accurate to within 10^{-14}?

   Solution: The error estimate is

   $$|f(x) - p_1(x)| \le \frac{1}{8}h^2 \max_{t \in [-1/2,1]} |-1/(1+t)^2| = \frac{1}{8}h^2(4) = h^2/2.$$

   So we require $h \le 0.1414213562 \times 10^{-6}$ for 10^{-14} accuracy; this means we need $n = (1/2)/h \approx 3.5 \times 10^6$ points. (Which is a *lot* of points!)

10. Construct a piecewise linear approximation to $f(x) = \sqrt{x}$ over the interval $[\frac{1}{4}, 1]$ using the nodes $\frac{1}{4}, \frac{9}{16}, 1$. What is the maximum error in this approximation?

11. Repeat the above for $f(x) = x^{1/3}$ over the interval $[\frac{1}{8}, 1]$, using the nodes $\frac{1}{8}, \frac{27}{64}, 1$.

    Solution: The polynomials are

    $$Q_1(x) = \frac{x - 1/8}{27/64 - 1/8}(3/4) + \frac{27/64 - x}{27/64 - 1/8}(1/2) = (32x + 15)/38,$$

    $$Q_2(x) = (16x + 21)/37.$$

    The maximum errors are given by

    $$|f(x) - Q_1(x)| \le \frac{1}{8}(19/64)^2 \max_{t \in [1/8,27/64]} |(-2/9)t^{-5/3}| = \frac{361}{32768} \times \frac{2}{9} \times 32 = \frac{361}{4608}$$

    $$= 0.7834201389 \times 10^{-1}$$

and

$$|f(x) - Q_2(x)| \leq \frac{1}{8}(37/64)^2 \max_{t \in [27/64,1]} |(-2/9)t^{-5/3}| = \frac{1369}{32768} \times \frac{2}{9} \times \frac{1024}{243} = \frac{1369}{34992}$$
$$= 0.3912322817 \times 10^{-1},$$

so the overall error bound is about 0.078.

◁ • • ▷

2.5 APPLICATION — THE TRAPEZOID RULE

Exercises:

1. Use the trapezoid rule with $h = \frac{1}{4}$ to approximate the integral

$$I = \int_0^1 x^3 dx = \frac{1}{4}.$$

   You should get that $T_4 = 17/64$. How small does h have to be to get that the error is less than 10^{-3}? 10^{-6}?

2. Use the trapezoid rule and $h = \pi/4$ to approximate the integral

$$I = \int_0^{\pi/2} \sin x \, dx = 1.$$

   How small does h have to be to get that the error is less than 10^{-3}? 10^{-6}?

3. Repeat the above with $h = \pi/5$.

4. Apply the trapezoid rule with $h = \frac{1}{8}$, to approximate the integral

$$I = \int_0^1 \frac{1}{\sqrt{1+x^4}} dx = 0.92703733865069.$$

   How small does h have to be to get that the error is less than 10^{-3}? 10^{-6}?

   Solution: It's a fairly direct (if tedious) computation to get

$$T_8 = \frac{1/8}{2} \left(\frac{1}{\sqrt{1+0^4}} + \frac{2}{\sqrt{1+(1/8)^4}} + \frac{2}{\sqrt{1+(2/8)^4}} + \frac{2}{\sqrt{1+(3/8)^4}} \right.$$
$$\left. + \frac{2}{\sqrt{1+(4/8)^4}} + \frac{2}{\sqrt{1+(5/8)^4}} + \frac{2}{\sqrt{1+(6/8)^4}} + \frac{2}{\sqrt{1+(7/8)^4}} + \frac{1}{\sqrt{1+1^4}} \right)$$
$$= \frac{1}{16} (1 + 1.99975590406939 + 1.99610515696578 + 1.98051315767607$$
$$+ 1.94028500029066 + 1.86291469895924 + 1.74315107424910$$
$$+ 1.58801110913806 + 0.70710678118655)$$
$$= 0.92611518015843.$$

Now, for the accuracy, we have

$$|I(f) - T_n(f)| = \frac{b-a}{12} h^2 |f''(\xi_h)|.$$

The second derivative is

$$f''(x) = \frac{6x^2(x^4 - 1)}{(1 + x^4)^{5/2}},$$

and we can bound this (for all $x \in [0, 1]$) as follows:

$$|f''(x)| \le |6(x^4 - 1)| \le 6.$$

A better upper bound would come from

$$|f''(x)| \le |6x^2(x^4 - 1)| \le 2.31$$

Therefore, to get 10^{-3} accuracy requires (using the weaker upper bound)

$$h \le \sqrt{(12/6) \times 10^{-3}} < 0.045.$$

To get 10^{-6} accuracy requires

$$h \le \sqrt{(12/6) \times 10^{-6}} < 0.00142.$$

5. Apply the trapezoid rule with $h = \frac{1}{8}$, to approximate the integral

$$I = \int_0^1 x(1 - x^2)dx = \frac{1}{4}.$$

Feel free to use a computer program or a calculator, as you wish. How small does h have to be to get that the error is less than 10^{-3}? 10^{-6}?

Solution: The approximation is $T_8(f) = 0.2460937500$. The error bound gives us

$$|I(f) - T_n(f)| \le (1/12)h^2 \max_{t \in [0,1]} |-6t| = h^2/2,$$

so we require $h \le 0.4472135955 \times 10^{-1}$ for 10^{-3} accuracy, and $h \le 0.1414213562 \times 10^{-2}$ for 10^{-6} accuracy.

6. Apply the trapezoid rule with $h = \frac{1}{8}$, to approximate the integral

$$I = \int_0^1 \ln(1 + x)dx = 2\ln 2 - 1.$$

How small does h have to be to get that the error is less than 10^{-3}? 10^{-6}?

Solution: The approximation is $T_8(f) = 0.3856439099$. The error bound yields

$$|I(f) - T_n(f)| \le (1/12)h^2 \max_{t \in [0,1]} |-1/(1 + t)^2| = h^2/12,$$

so we need $h \le 0.1095445115$ to get 10^{-3} accuracy, and $h \le 0.3464101615 \times 10^{-2}$ to get 10^{-6} accuracy.

7. Apply the trapezoid rule with $h = \frac{1}{8}$, to approximate the integral

$$I = \int_0^1 \frac{1}{1+x^3}\,dx = \frac{1}{3}\ln 2 + \frac{1}{9}\sqrt{3}\pi.$$

How small does h have to be to get that the error is less than 10^{-3}? 10^{-6}?

8. Apply the trapezoid rule with $h = \frac{1}{8}$, to approximate the integral

$$I = \int_1^2 e^{-x^2}\,dx = 0.1352572580.$$

How small does h have to be to get that the error is less than 10^{-3}? 10^{-6}?

Solution: We have

$$I_8 = \frac{1/8}{2}\left(e^{-1} + 2e^{-(81/64)} + 2e^{-(100/64)} + 2e^{-(121/64)} + 2e^{-(144/64)}\right.$$
$$\left. + 2e^{-(169/64)} + 2e^{-(196/64)} + 2e^{-(225/64)} + e^{-(256/64)}\right)$$

$$= \frac{1}{16}(0.56412590338763 + 0.41922277430220 + 0.30195483691183$$
$$+ 0.21079844912373 + 0.14263336539552 + 0.09354124476792$$
$$+ 0.05945843277232)$$
$$= 0.13612063042008.$$

To get the accuracy, we note that

$$f''(x) = (4x^2 - 2)e^{-x^2},$$

so that, over the interval $[1, 2]$,

$$|f''(x)| \le |4x^2 - 2|e^{-1} \le (16 + 2)e^{-1} = 18e^{-1} < 6.622.$$

Therefore, to get an error less than 10^{-3} we take h to satisfy

$$h \le \sqrt{(12/6.62) \times 10^{-3}} < 0.0426.$$

To get an error less than 10^{-6} we take h to satisfy

$$h \le \sqrt{(12/6.62) \times 10^{-6}} < 0.00135.$$

9. Let I_8 denote the value you obtained in the previous problem. Repeat the computation, this time using $h = \frac{1}{4}$, and call this approximate value I_4. Then compute

$$I_R = (4I_8 - I_4)/3$$

and compare this to the given exact value of the integral.

10. Repeat the above for the integral

$$I = \int_0^1 \frac{1}{1+x^3}\,dx = \frac{1}{3}\ln 2 + \frac{1}{9}\sqrt{3}\pi.$$

Solution: We get
$$I_4 = 0.8317002443,$$
$$I_8 = 0.8346696206,$$

and
$$I_R = 0.8356594123.$$

The exact value is $I(f) = 0.8356488485$, so we see that I_R is much more accurate than either of the trapezoid rule values.

11. For each integral below, write a program to do the trapezoid rule using the sequence of mesh sizes $h = \frac{1}{2}(b-a), \frac{1}{4}(b-a), \frac{1}{8}(b-a), \ldots, \frac{1}{128}(b-a)$, where $b - a$ is the length of the given interval. Verify that the expected rate of decrease of the error is obtained:

   (a) $f(x) = x^2 e^{-x}$, $[0,2]$, $I(f) = 2 - 10e^{-2} = 0.646647168$;

   (b) $f(x) = 1/(1 + 25x^2)$, $[0,1]$, $I(f) = \frac{1}{5}\arctan(5)$;

   (c) $f(x) = \sqrt{1 - x^2}$, $[-1,1]$, $I(f) = \pi/2$;

   (d) $f(x) = \ln x$, $[1,3]$, $I(f) = 3\ln 3 - 2 = 1.295836867$;

   (e) $f(x) = x^{5/2}$, $[0,1]$, $I(f) = 2/7$;

   (f) $f(x) = e^{-x}\sin(4x)$, $[0,\pi]$, $I(f) = \frac{4}{17}(1 - e^{-\pi}) = 0.2251261368$.

12. For each integral in Problem 11, how small does h have to be to get accuracy, according to the error theory, of at least 10^{-3}? 10^{-6}?

   Solution: For the single case of (d), since $f(x) = \ln x$, we have $f''(x) = -1/x^2$, so
$$|f''(x)| \leq 1$$

for all x on $[1,3]$. We therefore have
$$|I(f) - T_n(f)| \leq \frac{b-a}{12}h^2|f''(\xi_h)| \leq \frac{1}{6}h^2.$$

We therefore get 10^{-3} accuracy by imposing $\frac{1}{6}h^2 \leq 10^{-3}$ which implies $h \leq 0.0775$ For 10^{-6} accuracy we have $h \leq 0.00245$.

13. Apply the trapezoid rule to the integral
$$I = \int_0^1 \sqrt{x}\,dx = \frac{2}{3}$$

using a sequence of uniform grids with $h = \frac{1}{2}, \frac{1}{4}, \ldots$. Do we get the expected rate of convergence? Explain.

14. The length of a curve $y = g(x)$, for x between a and b, is given by the integral
$$L(g) = \int_a^b \sqrt{1 + [g'(x)]^2}\,dx.$$

Use the trapezoid rule with $h = \pi/4$ and $h = \pi/16$ to find the length of one "arch" of the sine curve.

Solution: We get
$$L(\sin) \approx 3.819943644$$

(using $h = \pi/4$), and
$$L(\sin) \approx 3.820197788$$

(using $h = \pi/16$).

15. Use the trapezoid rule to find the length of the logarithm curve between $a = 1$ and $b = e$, using $n = 4$ and $n = 16$.

16. What should h be to guarantee an accuracy of 10^{-8} when using the trapezoid rule for each of the following integrals:

(a)
$$I(f) = \int_0^1 e^{-x^2} dx;$$

Solution: The error bound gives us

$$|I(f) - T_n(f)| \le (1/12)h^2 \max_{t \in [0,1]} |(4t^2 - 2)e^{-t^2}| \le (h^2/12)(2)(1) = h^2/6,$$

so we require $h \le 0.2449489743 \times 10^{-3}$.

(b)
$$I(f) = \int_1^3 \ln x \, dx;$$

Solution: The error bound gives us

$$|I(f) - T_n(f)| \le (2/12)h^2 \max_{t \in [1,3]} |-1/t^2| \le (h^2/6)(1) = h^2/6$$

so we require $h \le 0.2449489743 \times 10^{-3}$.

(c)
$$I(f) = \int_{-5}^5 \frac{1}{1 + x^2} dx;$$

Solution: The error bound gives us

$$|I(f) - T_n(f)| \le (10/12)h^2 \max_{t \in [-5,5]} |(6x^2 - 2)/(x^2 + 1)^3| \le (5h^2/6)(2) = 5h^2/3$$

so we require $h \le 0.7745966692 \times 10^{-4}$.

(d)
$$I(f) = \int_0^1 \cos(\pi x/2) dx.$$

Solution: The error bound gives us

$$|I(f) - T_n(f)| \le (1/12)h^2 \max_{t \in [0,1]} |-(\pi/2)^2 \cos(\pi x/2)| \le \pi^2 h^2/48$$

so we require $h \le 0.2205315581 \times 10^{-3}$.

17. Since the natural logarithm is defined as an integral,

$$\ln x = \int_1^x \frac{1}{t} dt \qquad (2.3)$$

it is possible to use the trapezoid rule (or any other numerical integration rule) to construct approximations to $\ln x$.

(a) Show that using the trapezoid rule on the integral (2.3) results in the series approximation (for $x \in [1, 2]$)

$$\ln x \approx \frac{x^2 - 1}{2nx} + \sum_{k=1}^{n-1} \frac{x - 1}{n + k(x - 1)}.$$

Hint: What are a and b in the integral defining the logarithm?

(b) How many terms are needed in this approximation to get an error of less than 10^{-8} for all $x \in [1, 2]$? How many terms are needed for an error of less than 10^{-15} over the same interval?

(c) Implement this series for a predicted accuracy of 10^{-8} and compare it to the intrinsic natural logarithm function on your computer, over the interval $[1, 2]$. Is the expected accuracy achieved?

(d) If we were only interested in the interval $[1, 3/2]$, how many terms would be needed for the accuracy specified in (b)?

(e) Is it possible to reduce the computation of $\ln x$ for all $x > 0$ to the computation of $\ln z$ for $z \in [1, 3/2]$? Explain.

18. How small must h be to compute the error function,

$$\text{erf}(x) = \frac{2}{\sqrt{\pi}} \int_0^x e^{-t^2} dt,$$

using the trapezoid rule, to within 10^{-8} accuracy for all $x \in [0, 1]$?

Solution: The error estimate implies that

$$|\text{erf}(x) - T_n| \le (x/12)h^2 \max_{t \in [0,1]} |(2/\sqrt{\pi}(4t^2 - 2)e^{-t^2}| \le \frac{xh^2}{\sqrt{\pi}},$$

where $h = x/n$ and we have used a very crude upper bound on the second derivative. Since we want to achieve the specified accuracy for all $x \in [0, 1]$, we use $x = 1$ (since this maximizes the upper bound) and therefore solve the inequality

$$\frac{h^2}{\sqrt{\pi}} \le 10^{-8}$$

to get $h \le 0.1331335364 \times 10^{-3}$.

19. Use the data in Table 2.8 to compute

$$I = \int_1^2 \Gamma(x) dx$$

using $h = 0.2$ and $h = 0.1$.

20. Use the data in Table 2.7 to compute

$$I = \int_0^1 \mathrm{erf}(x)dx$$

using $h = 0.2$ and $h = 0.1$.

Solution: Using $h = 0.2$ we get

$$\int_0^1 \mathrm{erf}(x)dx \approx .4836804793;$$

using $h = 0.1$, we get

$$\int_0^1 \mathrm{erf}(x)dx \approx .4854701356.$$

21. Show that the trapezoid rule is *exact* for all linear polynomials.

22. Prove Theorem 2.4.

Solution: Let $h_i = x_i - x_{i-1}$; then we can apply Theorem 2.2 to get that

$$I(f) - T_n(f) = \sum_{i=1}^n \frac{-h_i^3}{12} f''(\xi_i),$$

where $\xi_i \in [x_{i-1}, x_i]$. Therefore,

$$
\begin{aligned}
|I(f) - T_n(f)| &\leq \left(\sum_{i=1}^n \frac{h_i^3}{12}\right) \max_{x \in [a,b]} |f''(x)| \\
&= \left(\sum_{i=1}^n \frac{h_i^2(x_i - x_{i-1})}{12}\right) \max_{x \in [a,b]} |f''(x)| \\
&\leq \frac{h^2}{12} \left(\sum_{i=1}^n (x_i - x_{i-1})\right) \max_{x \in [a,b]} |f''(x)| \\
&= \frac{h^2(b - a)}{12} \max_{x \in [a,b]} |f''(x)|.
\end{aligned}
$$

23. Extend the discussion on stability to include changes in the interval of integration instead of changes in the function. State and prove a theorem that bounds the change in the trapezoid rule approximation to

$$I(f) = \int_a^b f(x)dx$$

when the upper limit of integration changes from b to $b + \epsilon$, but f remains the same.

24. Consider a function $\tilde{f}(x)$ which is the floating point representation of $f(x)$; thus $f - \tilde{f}$ is the rounding error in computing f. If we assume that $|f(x) - \tilde{f}(x)| \leq \epsilon$ for all x, show that

$$|T_n(f) - T_n(\tilde{f})| \leq \epsilon(b - a).$$

What does this say about the effects of rounding error on the trapezoid rule?

Solution:

$$T_n(f) - T_n(\tilde{f}) = (h/2) \sum_{i=1}^{n} \left(f(x_i) - \tilde{f}(x_i) + f(x_{i-1}) - \tilde{f}(x_{i-1}) \right),$$

so

$$|T_n(f) - T_n(\tilde{f})| \le (h/2) \sum_{i=1}^{n} \left(e(x_i) + e(x_{i-1}) \right),$$

where $e(x) = f(x) - \tilde{f}(x)$. The result follows by simple manipulation.

25. We will end this section with a few exercises designed to illustrate the `trapz` command. Strictly speaking, `trapz` simply does a specialized summation. If S = `trapz(Y)`, then

$$S = \sum_{i=0}^{n} {''} Y_i = \frac{1}{2}Y_0 + Y_1 + Y_2 + \ldots + Y_{n-1} + \frac{1}{2}Y_n,$$

which is the trapezoid rule with $h = 1$. To do an actual trapezoid rule for a uniform discretization, just multiply by h; for $h = \frac{b-a}{n}$, we have

$$\int_a^b f(x)dx \approx h \sum_{i=0}^{n} {''} f(x_i) = T_n(f) = \text{h} * \text{trapz(f)},$$

where f is the vector of $\{f(x_i)\}_{i=0}^{n}$ values.

For each of the following integrals, use `trapz` to compute an approximate value.

(a) Apply the trapezoid rule with $h = \frac{1}{8}$ to approximate the integral

$$I = \int_0^1 \ln(1 + x)dx = 2\ln 2 - 1.$$

Compare what you got here to what you got in Problem 6.
Solution: The following script will do the job. I got `trapy` = 0.385643909952095; in Problem 6 we got 0.3856439099.

```
N = 8;
h = 1/N;
x = [0:N]/N;
y = log(1 + x);
S = trapz(y);
trapy = h*S
```

(b) Repeat the above for $h = 0.01$.
Solution: If we set N = 100 in the above script, we get

$$\text{trapy} = 0.386290194477529,$$

and the error is now much less than before.

(c) Apply the trapezoid rule with $h = \frac{1}{8}$ to approximate the integral

$$I = \int_0^1 \frac{1}{1+x^3} dx = \frac{1}{3} \ln 2 + \frac{1}{9}\sqrt{3}\pi.$$

Compare what you got here to what you got in Problem 7.

(d) Repeat the above for $h = 0.01$.

(e) Apply the trapezoid rule with $h = \frac{1}{8}$ to approximate the integral

$$I = \int_1^2 e^{-x^2} dx = 0.1352572580.$$

Compare what you got here to what you got in Problem 8.

(f) Repeat the above for $h = 0.01$.

2.6 SOLUTION OF TRIDIAGONAL LINEAR SYSTEMS

Exercises:

1. Use the tridiagonal algorithm in this section to compute the solution to the following system of equations:

$$\begin{bmatrix} 4 & 2 & 0 & 0 \\ 1 & 4 & 1 & 0 \\ 0 & 1 & 4 & 1 \\ 0 & 0 & 2 & 4 \end{bmatrix} \begin{bmatrix} x_1 \\ x_2 \\ x_3 \\ x_4 \end{bmatrix} = \begin{bmatrix} \pi/9 \\ \sqrt{3}/2 \\ \sqrt{3}/2 \\ -\pi/9 \end{bmatrix}.$$

Solution: After the elimination step is completed, the triangular system is

$$\begin{bmatrix} 4 & 2 & 0 & 0 \\ 0 & 3.5 & 1 & 0 \\ 0 & 0 & 3.7143 & 1 \\ 0 & 0 & 0 & 3.4615 \end{bmatrix} \begin{bmatrix} x_1 \\ x_2 \\ x_3 \\ x_4 \end{bmatrix} = \begin{bmatrix} 0.3491 \\ 0.7788 \\ 0.6435 \\ -0.6956 \end{bmatrix}.$$

and the solution is

$$\begin{bmatrix} x_1 \\ x_2 \\ x_3 \\ x_4 \end{bmatrix} = \begin{bmatrix} 0.0085 \\ 0.1575 \\ 0.2274 \\ -0.2009 \end{bmatrix}.$$

2. Write a computer code to solve the previous problem

   Solution: The following is a MATLAB script that produced the previous solution

```
function [delta, f, x] = trisol(l,d,u,b)
    n = length(d);
    x = zeros(n,1);
    for k=2:n
        d(k) = d(k) - u(k-1)*l(k)/d(k-1);
        b(k) = b(k) - b(k-1)*l(k)/d(k-1);
    end
    x(n) = b(n)/d(n);
    for k=(n-1):(-1):1
        x(k) = (b(k) - u(k)*x(k+1))/d(k);
    end
    delta = d;
    f = b;
```

3. Use the algorithm of this section to solve the following system of equations:

$$\begin{bmatrix} 6 & 1 & 0 & 0 \\ 2 & 4 & 1 & 0 \\ 0 & 1 & 4 & 2 \\ 0 & 0 & 1 & 6 \end{bmatrix} \begin{bmatrix} x_1 \\ x_2 \\ x_3 \\ x_4 \end{bmatrix} = \begin{bmatrix} 8 \\ 13 \\ 22 \\ 27 \end{bmatrix}.$$

   You should get the solution $x = (1, 2, 3, 4)^T$.

4. Write a computer code to solve the previous problem.

5. The diagonal dominance condition is an example of a *sufficient* but not *necessary* condition. That is, the algorithm will often work for systems that are *not* diagonally dominant. Show that the following system is not diagonally dominant, but then use the tridiagonal algorithm in this section to compute the solution to it:

$$\begin{bmatrix} 1 & \frac{1}{2} & 0 & 0 \\ \frac{1}{2} & \frac{1}{3} & \frac{1}{4} & 0 \\ 0 & \frac{1}{4} & \frac{1}{5} & \frac{1}{6} \\ 0 & 0 & \frac{1}{6} & \frac{1}{7} \end{bmatrix} \begin{bmatrix} x_1 \\ x_2 \\ x_3 \\ x_4 \end{bmatrix} = \begin{bmatrix} 2 \\ 23/12 \\ 53/30 \\ 15/14 \end{bmatrix}.$$

   You should get the solution $x = (1, 2, 3, 4)^T$.

   Solution: The matrix fails to be diagonally dominant because of the values in the second, third, and fourth rows. However, if we apply the algorithm, we get the triangular system

$$\begin{bmatrix} 1 & \frac{1}{2} & 0 & 0 \\ 0 & 0.0833 & \frac{1}{4} & 0 \\ 0 & 0 & -0.5500 & \frac{1}{6} \\ 0 & 0 & 0 & 0.1934 \end{bmatrix} \begin{bmatrix} x_1 \\ x_2 \\ x_3 \\ x_4 \end{bmatrix} = \begin{bmatrix} 2.0000 \\ 0.9167 \\ -0.9833 \\ 0.7734 \end{bmatrix},$$

   and then the correct values for the solution, x. (It always helps to type in the correct right-side vector when checking these things.)

6. Use the tridiagonal algorithm in this section to compute the solution to the following system of equations:

$$
\begin{bmatrix}
1 & \frac{1}{2} & 0 & 0 \\
\frac{1}{2} & \frac{1}{3} & \frac{1}{4} & 0 \\
0 & \frac{1}{4} & \frac{1}{5} & \frac{1}{6} \\
0 & 0 & \frac{1}{6} & \frac{1}{7}
\end{bmatrix}
\begin{bmatrix}
x_1 \\ x_2 \\ x_3 \\ x_4
\end{bmatrix}
=
\begin{bmatrix}
2 \\ 2 \\ 53/30 \\ 15/14
\end{bmatrix}.
$$

Note that this is a very small change from the previous problem, since the only difference is that b_2 has changed by only $1/12$. How much has the answer changed?

Solution: The solution is now

$$
\begin{bmatrix}
x_1 \\ x_2 \\ x_3 \\ x_4
\end{bmatrix}
=
\begin{bmatrix}
1.0037 \\ 1.9925 \\ 3.3358 \\ 3.6082
\end{bmatrix}.
$$

Considering the small change in the problem, this is a substantial change in the solution.

7. Write a computer code to do the previous two problems.

8. Verify that the following system is diagonally dominant and use the algorithm of this section to find the solution.

$$
\begin{bmatrix}
\frac{1}{2} & \frac{10}{21} & 0 & 0 \\
\frac{1}{4} & \frac{1}{3} & \frac{1}{13} & 0 \\
0 & \frac{1}{5} & \frac{1}{4} & \frac{1}{21} \\
0 & 0 & \frac{1}{6} & \frac{1}{5}
\end{bmatrix}
\begin{bmatrix}
x_1 \\ x_2 \\ x_3 \\ x_4
\end{bmatrix}
=
\begin{bmatrix}
\frac{61}{42} \\ \frac{179}{156} \\ \frac{563}{420} \\ \frac{13}{10}
\end{bmatrix}.
$$

9. Use the algorithm of this section to find the solution to this system:

$$
\begin{bmatrix}
\frac{1}{2} & \frac{10}{21} & 0 & 0 \\
\frac{1}{4} & \frac{1}{3} & \frac{1}{13} & 0 \\
0 & \frac{1}{5} & \frac{1}{4} & \frac{1}{21} \\
0 & 0 & \frac{1}{6} & \frac{1}{5}
\end{bmatrix}
\begin{bmatrix}
x_1 \\ x_2 \\ x_3 \\ x_4
\end{bmatrix}
=
\begin{bmatrix}
\frac{61}{42} \\ \frac{180}{156} \\ \frac{563}{420} \\ \frac{13}{10}
\end{bmatrix}.
$$

Note that the right side here is different from that in the previous problem by only a small amount in the b_2 component. Comment on your results here as compared to those in the previous problem.

Solution: The solution is now

$$
\begin{bmatrix}
x_1 \\ x_2 \\ x_3 \\ x_4
\end{bmatrix}
=
\begin{bmatrix}
0.7236 \\ 2.2902 \\ 2.7240 \\ 4.2300
\end{bmatrix}.
$$

Again, this is a very large change in the solution for so modest a change in the problem.

10. Write a computer code to do the previous two problems.

11. Write a code that carries out the tridiagonal solution algorithm, and test it on the following system of equations,

$$Tx = b$$

where T is 10×10 with

$$t_{ij} = \begin{cases} 1, & |i - j| = 1; \\ j + 1, & i = j; \\ 0, & \text{otherwise}; \end{cases}$$

and $b_i = 1$ for all i. Check your results by computing the residual $r = b - Tx$. What is the largest component (in absolute value) of r? (You could also check your results by using MATLAB's backslash operator to solve the system.)

12. Extend the tridiagonal algorithm to a *pentadiagonal* matrix, i.e., one with five non-zero diagonals. Write a program to carry out this solution algorithm, and apply it to the system

$$\begin{bmatrix} 4 & 2 & 1 & 0 \\ 1 & 4 & 1 & 1 \\ 1 & 1 & 4 & 1 \\ 0 & 1 & 2 & 4 \end{bmatrix} \begin{bmatrix} x_1 \\ x_2 \\ x_3 \\ x_4 \end{bmatrix} = \begin{bmatrix} 1 \\ 1 \\ 1 \\ 1 \end{bmatrix}.$$

Check your results by again computing the residual vector, or by using MATLAB's backslash operation.

Solution: A MATLAB script for doing this is given below. On the example in the exercise it returns the (exact) solution $x = (1/7, 1/7, 1/7, 1/7)^T$.

```
function [delta, f, x] = pentasol(k,l,d,u,v,b)
    n = length(d);
    x = zeros(n,1);
    d(2) = d(2) - l(2)*u(1)/d(1);
    u(2) = u(2) - l(2)*v(1)/d(1);
    b(2) = b(2) - l(2)*b(1)/d(1);
    for j=3:n
        l(j) = l(j) - k(3)*u(j-2)/d(j-2);
        d(j) = d(j) - k(j)*v(j-2)/d(j-2);
        b(j) = b(j) - k(j)*b(j-2)/d(j-2);
        d(j) = d(j) - l(j)*u(j-1)/d(j-1);
        if j < n
            u(j) = u(j) - l(j)*v(j-1)/d(j-1);
        end
        b(j) = b(j) - l(j)*b(j-1)/d(j-1);
    end
    delta = d;
    f = b;
    x = f;
%
    x(n) = f(n)/d(n);
    x(n-1) = (b(n-1) - u(n-1)*x(n))/d(n-1);
    for j=(n-2):(-1):1
        x(j) = (b(j) - u(j)*x(j+1) - v(j)*x(j+2))/d(j)
```

```
end
```

13. Consider the family of tridiagonal problems defined by the matrix $K_n \in \mathbb{R}^{n \times n}$, with

$$K_n = \text{tridiag}(-1, 2, -1)$$

and a randomly defined right-hand-side vector. (Use `rand` to generate the random vectors.) Solve the system over the range $4 \leq n \leq 10,000$; use the appropriate timing functions to estimate the CPU time required for each case, and plot the result as a function of n. Comment on your results. (You will probably want to use a semilog plot.)

Solution: I got the plot in Fig. 2.5. While this is a "noisy" plot, it is very much generally a logarithmic curve, which suggests the actual timing numbers are a straight line, which is what we would expect, given that the cost of doing the solution is linear in the size of the matrix.

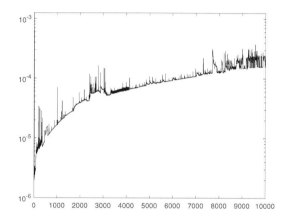

Figure 2.5   Estimated timing cost for Problem 13, semilog scale.

◁ ● ● ▷

2.7   APPLICATION: SIMPLE TWO-POINT BOUNDARY VALUE PROBLEMS

Exercises:

1. Solve, by hand, the two-point BVP (2.28) and (2.29) when $f(x) = x$, using $h = \frac{1}{4}$. Write out the linear system explicitly prior to solution. You should get the following 3×3 system:

$$
\begin{bmatrix}
2.0625 & -1 & 0 \\
-1 & 2.0625 & -1 \\
0 & -1 & 2.0625
\end{bmatrix}
\begin{bmatrix}
U_1 \\
U_2 \\
U_3
\end{bmatrix}
=
\begin{bmatrix}
0.015625 \\
0.03125 \\
0.046875
\end{bmatrix}.
$$

Solution: The approximate solution is

$$
\begin{bmatrix} U_1 \\ U_2 \\ U_3 \end{bmatrix} = \begin{bmatrix} 0.0349 \\ 0.0563 \\ 0.0500 \end{bmatrix}.
$$

2. Repeat the above, this time using $h = \frac{1}{5}$. What is the system now?

3. Repeat it again, this time using $h = \frac{1}{8}$. What is the system now? (For this problem, you probably will want to use a computer code to actually solve the system.)

Solution: The approximate solution is

$$
\begin{bmatrix} U_1 \\ U_2 \\ U_3 \\ U_4 \\ U_5 \\ U_6 \\ U_7 \end{bmatrix} = \begin{bmatrix} 0.0173 \\ 0.0331 \\ 0.0459 \\ 0.0540 \\ 0.0556 \\ 0.0486 \\ 0.0309 \end{bmatrix}.
$$

4. Write a program which solves the two-point BVP (2.28) and (2.29) where f is as given below:

(a) $f(x) = 4e^{-x} - 4xe^{-x}$, $\quad u(x) = x(1-x)e^{-x}$;

(b) $f(x) = (\pi^2 + 1)\sin\pi x$, $\quad u(x) = \sin\pi x$;

(c) $f(x) = \pi(\pi\sin\pi x + 2\cos\pi x)e^{-x}$, $\quad u(x) = e^{-x}\sin\pi x$;

(d) $f(x) = 3 - \frac{1}{x} - (x^2 - x - 2)\log x$, $\quad u(x) = x(1-x)\log x$.

The exact solutions are as given. Using $h^{-1} = 4, 8, 16, \ldots$, do we get the same kind of accuracy as in Table 2.10? Explain why or why not.

Solution: The approximate solutions will display the kind of $\mathcal{O}(h^2)$ accuracy suggested in the text, *except* for (d); in this case, the singularity in the logarithm function in the solution affects the accuracy of the approximation.

5. Try to apply the ideas of this section to approximating the solution of the two point boundary value problem

$$
\begin{aligned}
-u'' + u' + u &= 1, \quad x \in [0, 1] \\
u(0) = 0, u(1) &= 0.
\end{aligned}
$$

Can we get a tridiagonal system that Algorithm 2.6 can be applied to? Hint: Consider some of the approximations from §2.2; use the most accurate ones that can be easily used.

Solution: Use the approximation

$$
u'(x) \approx \frac{u(x+h) - u(x-h)}{2h}
$$

to construct the approximation. Using the approximation

$$
u'(x) \approx \frac{u(x+h) - u(x)}{h}
$$

will result in a loss of accuracy.

6. Solve the two-point boundary value problem problem

$$-u'' + 64u' + u = 1, \quad x \in [0, 1],$$
$$u(0) = 0, u(1) = 0,$$

using a range of mesh sizes, starting with $h = 1/4, 1/8$, and going as far as $h = 1/256$. Comment on your results.

Solution: For larger values of h, the approximate solution is erratic and wildly oscillatory. It isn't until about $h \leq 32$ or so that the solution begins to settle down.

7. Generalize the solution of the two-point boundary value problem to the case where $u(0) = g_0 \neq 0$ and $u(1) = g_1 \neq 0$. Apply this to the solution of the problem

$$-u'' + u = 0, \quad x \in [0, 1],$$
$$u(0) = 2, u(1) = 1,$$

which has exact solution

$$u(x) = \left(\frac{e - 2}{e^2 - 1}\right) e^x + \left(\frac{2e - 1}{e^2 - 1}\right) e^{-x}.$$

Solve this for a range of values of the mesh. Do we get the expected $\mathcal{O}(h^2)$ accuracy?

8. Consider the problem of determining the deflection of a thin beam, supported at both ends, due to a uniform load being placed along the beam. In one simple model, the deflection $u(x)$ as a function of position x along the beam satisfies the boundary value problem

$$-u'' + pu = -qx(L - x), \quad 0 < x < L;$$
$$u(0) = u(L) = 0.$$

Here p is a constant that depends on the material properties of the beam, L is the length of the beam, and q depends on the material properties of the beam as well as the size of the load placed on the beam. For a six-foot-long beam, with $p = 7 \times 10^{-6}$ and $q = 4 \times 10^{-7}$, what is the maximum deflection of the beam? Use a fine enough grid that you can be confident of the accuracy of your results. Note that this problem is slightly more general than our example (2.28)-(2.29); you will have to adapt our method to this more general case.

Solution: See Figure 2.6.

9. Repeat the above problem, except this time use a three-foot-long beam. How much does the maximum deflection change, and is it larger or smaller?

10. Repeat the beam problem again, but this time use a 12-foot-long beam.

11. Try to apply the ideas of this section to the solution of the *nonlinear* boundary value problem defined by

$$-u'' + e^u = 0, \quad 0 < x < 1;$$
$$u(0) = u(1) = 0.$$

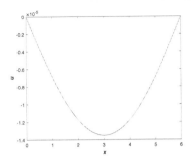

Figure 2.6   Exact solution for Exercise 2.7.8.

Write out the systems of equations for the specific case of $h = \frac{1}{4}$. What goes wrong? Why can't we proceed with the approximate solution?

Solution: The problem is that the system of equations that is produced by the discretization process is *nonlinear*, so our inherently linear algorithm won't work.

◁ • • ▷

CHAPTER 3

ROOT-FINDING

3.1 THE BISECTION METHOD

Exercises:

1. Do three iterations (by hand) of the bisection method, applied to $f(x) = x^3 - 2$, using $a = 0$ and $b = 2$.

   Solution: We have $f(a) = -2$, $f(b) = 6$. The first iteration gives us

   $$c = 1, \quad f(c) = -1,$$

   so the new interval is $[a, b] = [1, 2]$. The second iteration then gives us

   $$c = 3/2, \quad f(c) = 11/8,$$

   so the new interval is $[a, b] = [1, 3/2]$. Finally, the third iteration then gives us

   $$c = 5/4, \quad f(c) = -3/64,$$

   so the new interval is $[a, b] = [5/4, 3/2]$.

2. For each of the functions listed below, do a calculation by hand (i.e., with a calculator) to find the root to an accuracy of 0.1. This will take at most 5 iterations for all of these, and fewer for several of them.

(a) $f(x) = x - e^{-x^2}$, $[a, b] = [0, 1]$;

(b) $f(x) = \ln x + x$, $[a, b] = [\frac{1}{10}, 1]$;

(c) $f(x) = x^3 - 3$, $[a, b] = [0, 3]$;

(d) $f(x) = x^6 - x - 1$, $[a, b] = [0, 2]$;

(e) $f(x) = 3 - 2^x$, $[a, b] = [0, 2]$.

3. Write a program which uses the bisection method to find the root of a given function on a given interval, and apply this program to find the roots of the functions below on the indicated intervals. Use the relationship (3.2) to determine *a priori* the number of steps necessary for the root to be accurate to within 10^{-6}:

(a) $f(x) = x^3 - 2$, $[a, b] = [0, 2]$;

(b) $f(x) = e^x - 2$, $[a, b] = [0, 1]$;

(c) $f(x) = x - e^{-x}$, $[a, b] = [0, 1]$;

(d) $f(x) = x^6 - x - 1$, $[a, b] = [0, 2]$;

(e) $f(x) = x^3 - 2x - 5$, $[a, b] = [0, 3]$;

(f) $f(x) = 1 - 2xe^{-x/2}$, $[a, b] = [0, 2]$;

(g) $f(x) = 5 - x^{-1}$, $[a, b] = [0.1, 0.25]$;

(h) $f(x) = x^2 - \sin x$, $[a, b] = [0.1, \pi]$.

Solution: The solutions for some of the functions in this list are given in Table 3.1. For the sake of brevity, we give only the number of iterations required plus the final answer.

Table 3.1   Solutions to Problem 3.1.3

$f(x)$	n	c
$e^x - 2$	20	0.69314670562744
$x^3 - 2x - 5$	22	2.09455132484436
$x^2 - \sin x$	22	0.87672686579500

4. Use the bisection algorithm to solve the nonlinear equation $x = \cos x$. Choose your own initial interval by some judicious experimentation with a calculator.

5. Use the bisection algorithm to solve the nonlinear equation $x = \exp^{-x}$. Again, choose your own initial interval by some judicious experimentation with a calculator.

6. If you borrow L dollars at an annual interest rate of r, for a period of m years, then the size of the monthly payment, M, is given by the *annuity equation*,

$$L = \frac{12M}{r}[1 - (1 + (r/12)^{-12m})].$$

The author needs to borrow \$150,000 to buy the new house that he wants, and he can only afford to pay \$600 per month. Assuming a 30 year mortgage, use the bisection method to determine what interest rate he can afford to pay. (Should the author perhaps find some rich relatives to help him out here?)

Solution: This amounts to a root-finding problem for

$$f(r) = L - \frac{12M}{r}[1 - (1 + (r/12)^{-12m})],$$

where $L = 150$, $M = 0.6$, and $m = 30$. The interest rate turns out to be $r = 0.02593381500244$, or, about 2.6%. The author is already checking for a rich relative.

7. What is the interest rate that the author can afford if he only has to borrow \$100,000?

8. Consider the problem of modeling the position of the liquid-solid boundary in a substance that is melting due to the application of heat at one end. In a simplified model,[1] if the initial position of the interface is taken to be $x = 0$, then the interface moves according to

$$x = 2\beta\sqrt{t},$$

where β satisfies the nonlinear equation

$$\frac{(T_M - T_0)k}{\lambda\sqrt{\pi}}e^{-\beta^2/k} = \beta\text{erf}(\beta/\sqrt{k}).$$

Here T_M is the melting temperature (absolute scale), $T_0 < T_M$ is the applied temperature, k and λ are parameters dependent on the material properties of the substance involved, and $\text{erf}(z)$ is the *error function* defined by

$$\text{erf}(z) = \frac{2}{\sqrt{\pi}} \int_0^z e^{-t^2} dt.$$

MATLAB has an intrinsic error function, `erf`. If you are not using MATLAB, the error function can be accurately approximated by

$$E(z) = 1 - (a_1\xi + a_2\xi^2 + a_3\xi^3)e^{-z^2},$$

where $\xi = 1/(1 + pz)$ and

$$p = 0.47047, \ a_1 = 0.3480242, \ a_2 = -0.0958798, \ a_3 = 0.747856.$$

(a) Show that finding β is equivalent to finding the root α of the one-parameter family of functions defined by

$$f(z) = \theta e^{-z^2} - z\text{erf}(z). \tag{3.1}$$

What is θ? How is α related to β?

(b) Find the value of α corresponding to $\theta = 0.001, 0.1, 10, 1000$. Use the bisection method, and $E(z)$ to approximate the error function.

Solution:

(a) Make the change of variable $z = \beta/\sqrt{k}$, $\theta = (T_M - T_0)k/\lambda\sqrt{\pi}$.

(b) For $\theta = 0.001$, $\alpha = 0.0297$; for $\theta = 0.1$, $\alpha = 0.2895$; for $\theta = 10$, $\alpha = 1.4149$; for $\theta = 1000$, $\alpha = 2.4519$.

[1] See L.I. Rubinstein, *The Stefan Problem*, American Mathematical Society, Providence, RI, 1971.

9. A variation on the bisection method is known as *regula falsi*, or, *the method of false position*. Given an interval $[a, b]$, with $f(a)f(b) < 0$, the new point c is defined by finding where the straight line connecting $(a, f(a))$ and $(b, f(b))$ crosses the axis. Show that this yields

$$c = b - f(b)(b - a)/(f(b) - f(a)).$$

10. Do three iterations (by hand) of regula-falsi (see Problem 9), applied to $f(x) = x^3 - 2$, using $a = 0$ and $b = 2$. Compare to your results for Problem 1.

    Solution: Regula-falsi produces the results shown in Table 3.2. Note that after three iterations, the interval is reduced to $[1.0696, 2]$, whereas bisection would have reduced the interval to $[1.25, 1.50]$, which is shorter. Thus bisection appears to be superior to regula-falsi, at least on this example.

Table 3.2 Solutions to Problem 3.1.10.

c	$f(c)$	a	b
0.5000	−1.8750	0.5000	2.0000
0.8571	−1.3703	0.8571	2.0000
1.0696	−0.7763	1.0696	2.0000

11. Modify the bisection algorithm to perform regula falsi (see Problem 9), and use the new method to find the same roots as in Problem 3. Stop the program when the difference between consecutive iterates is less than 10^{-6}, i.e., when $|x_{k+1} - x_k| \leq 10^{-6}$, or when the function value satisfies $|f(c_k)| \leq 10^{-6}$.

12. Repeat Problem 8(b), using your regula-falsi program.

    Solution: Regula-falsi finds the same values as bisection, but more slowly.

13. The bisection method will always cut the interval of uncertainty in half, but regula falsi might cut the interval by less, or might cut it by more. Do both bisection and regula falsi on the function $f(x) = e^{-4x} - \frac{1}{10}$, using the initial interval $[0, 5]$. Which one gets to the root the fastest?

14. Apply both bisection and regula-falsi to the following functions on the indicated intervals. Comment on your results in the light of how the methods are supposed to behave:

    (a) $f(x) = 1/(x - 1)$, $[a, b] = [0, 3]$;
    (b) $f(x) = 1/(x^2 + 1)$, $[a, b] = [0, 5]$.

    Solution: This is kind of a trick question. Neither function has a root on the interval in question, but both methods will try to find it. Regula falsi lands on the singularity for $f(x) = 1/(x - 1)$, but bisection doesn't, so keeps happily computing away.

◁ • • ▷

3.2 NEWTON'S METHOD: DERIVATION AND EXAMPLES

Exercises:

1. Write down Newton's method as applied to the function $f(x) = x^2 - 2$. Simplify the computation as much as possible. What has been accomplished if we find the root of this function?

2. Write down Newton's method as applied to the function $f(x) = x^3 - 2$. Simplify the computation as much as possible. What has been accomplished if we find the root of this function?

   Solution:

$$
\begin{aligned}
x_{n+1} &= x_n - \frac{f(x_n)}{f'(x_n)} \\
&= x_n - \left(\frac{x_n^3 - 2}{3x_n^2} \right) \\
&= (2/3)(x_n + (1/x_n^2)).
\end{aligned}
$$

   When we find the root of this function, we have found $2^{1/3}$, the cube root of 2.

3. Generalize the preceding two problems by writing down Newton's method as applied to $f(x) = x^n - a$.

   Solution:

$$
\begin{aligned}
x_{n+1} &= x_n - \frac{f(x_n)}{f'(x_n)} \\
&= x_n - \left(\frac{x_n^N - a}{N x_n^{N-1}} \right) \\
&= (1/N)((N-1)x_n + (1/x_n^{N-1})).
\end{aligned}
$$

   When we find the root of this function, we have found $a^{1/N}$, the Nth root of a.

4. Write down Newton's method as applied to the function $f(x) = a - x^{-1}$. Simplify the resulting computation as much as possible. What has been accomplished if we find the root of this function?

5. Do three iterations of Newton's method (by hand) for each of the following functions:

   (a) $f(x) = x^6 - x - 1$, $x_0 = 1$;

   Solution: We get:

$$
x_1 = 1 - f(1)/f'(1) = 1 - \frac{-1}{5} = 1.2;
$$

$$
x_2 = 1.2 - f(1.2)/f'(1.2) = 1 - \frac{0.785984}{13.92992} = 1.143575843;
$$

$$
\begin{aligned}
x_3 &= 1.143575843 - f(1.143575843)/f'(1.143575843) \\
&= 1.143575843 - \frac{0.93031963}{10.73481139} \\
&= 1.134909462.
\end{aligned}
$$

(b) $f(x) = x + \tan x$, $x_0 = 3$;

(c) $f(x) = 1 - 2xe^{-x/2}$, $x_0 = 0$;

(d) $f(x) = 5 - x^{-1}$, $x_0 = \frac{1}{2}$;

(e) $f(x) = x^2 - \sin x$, $x_0 = \frac{1}{2}$;

(f) $f(x) = x^3 - 2x - 5$, $x_0 = 2$.

Solution: We get:

$$x_1 = 2 - f(2)/f'(2) = 2.1;$$

$$x_2 = 2.1 - f(2.1)/f'(2.1) = 2.094568121;$$

$$x_3 = 2.094568121 - f(2.094568121)/f'(2.094568121) = 2.094551482.$$

(g) $f(x) = e^x - 2$, $x_0 = 1$;

Solution: We get:

$$x_1 = 1 - f(1)/f'(1) = 0.7357588825;$$

$$x_2 = 0.7357588825 - f(0.7357588825)/f'(1.7357588825) = 0.6940422999;$$

$$x_3 = 0.6940422999 - f(0.6940422999)/f'(0.6940422999) = 0.6931475811.$$

(h) $f(x) = x^3 - 2$, $x_0 = 1$;

Solution: We get:

$$x_1 = 1 - f(1)/f'(1) = 1.333333333;$$

$$x_2 = 1.333333333 - f(1.333333333)/f'(1.333333333) = 1.263888889;$$

$$x_3 = 1.263888889 - f(1.263888889)/f'(1.263888889) = 1.259933493.$$

(i) $f(x) = x - e^{-x}$, $x_0 = 1$;

Solution: We get:

$$x_1 = 1 - f(1)/f'(1) = 0.5378828427;$$

$$x_2 = 0.5378828427 - f(0.5378828427)/f'(0.5378828427) = 0.5669869914;$$

$$x_3 = 0.5669869914 - f(0.5669869914)/f'(0.5669869914) = 0.5671432860.$$

(j) $f(x) = 2 - x^{-1} \ln x$, $x_0 = \frac{1}{3}$.

6. Do three iterations of Newton's method for $f(x) = 3 - e^x$, using $x_0 = 1$. Repeat, using $x_0 = 2, 4, 8, 16$. Comment on your results.

7. Draw the graph of a single function f which satisfies all of the following:

   (a) f is defined and differentiable for all x;

   (b) There is a unique root $\alpha > 0$;

   (c) Newton's method will converge for any $x_0 > \alpha$;

   (d) Newton's method will diverge, or converge to $x_* \neq \alpha$, for all $x_0 < 0$.

   Note that the requirement here is to find a single function that satisfies all of these conditions.

   Solution: Figure 3.1 shows a possible solution.

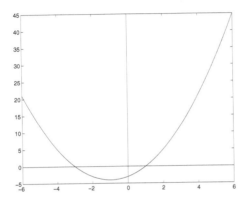

   Figure 3.1   Solution for Problem 3.2.7.

8. Draw the graph of a single function f which satisfies all of the following:

   (a) f is defined and differentiable for all x;

   (b) There is a single root $\alpha \in [a, b]$, for some interval $[a, b]$;

   (c) There is a second root, $\beta < a$;

   (d) Newton's method will converge to α for all $x_0 \in [a, b]$;

   (e) Newton's method will converge to β for all $x_0 \leq \beta$.

   Note that the requirement here, again, is to find a single function that satisfies all of these conditions.

9. Write down Newton's method for finding the root of the arctangent function. From this formulate an equation which must be satisfied by the value $x = \beta$, in order to have the Newton iteration cycle back and forth between β and $-\beta$. Hint: If $x_n = \beta$, and Newton's method is supposed to give us $x_{n+1} = -\beta$, what is an equation satisfied by β?

   Solution: Newton's method applied to the arctangent function is

   $$x_{n+1} = x_n - (x_n^2 + 1)(\arctan x_n).$$

What we want to happen is that $x_0 = \beta$ leads to $x_1 = -\beta$. This means we have

$$-\beta = \beta - (\beta^2 + 1)(\arctan \beta),$$

or,

$$2\beta - (\beta^2 + 1)(\arctan \beta) = 0.$$

Graphing this function shows three roots: one at $\beta = 0$, and two at $\beta \approx \pm 1.39$. We want one of these last two.

10. Compute the value of β from the previous problem. Hint: Use bisection to solve the equation for β.

11. Use Newton's method on the computer of your choice to compute the root $\alpha = 0$ of the arctangent function. Use the value of β from the previous problem as your x_0 and comment on your results. Repeat using $x_0 = \beta/2$ and $x_0 = \beta - \epsilon$, where ϵ is $\mathcal{O}(\mathbf{u})$.

Solution: Table 3.3 shows the first five iterates using the value of β found in Problem 10. Note that the iteration is not *exactly* cycling back and forth between β and $-\beta$. This is because of rounding error in the computation. If we use $x_0 = \beta/2$, then the computation converges to the root at $x = 0$ very quickly: $x_3 \approx 10^{-7}$. Taking x_0 closer to β only slows down this convergence. For example, taking $x_0 = (0.999)\beta$ yields $x_5 \approx 0.36$.

Table 3.3   Solution to Problem 3.2.11.

k	x_k
1	-1.39174467908955
2	1.39174382537168
3	-1.39174157322777
4	1.39173563198669
5	-1.39171995883655

◁ • • ▷

3.3   HOW TO STOP NEWTON'S METHOD

Exercises:

1. Under the assumption that $f'(\alpha) \neq 0$ and $x_n \to \alpha$, prove (3.10); be sure to provide all the details. Hint: Expand f and f' in Taylor series about $x = \alpha$.

   Solution: We have

   $$
   \begin{aligned}
   C_n &= \left(\frac{f(x_n)}{f(x_{n-1})} \right) \left(\frac{f'(x_{n-1})}{f'(c_n)} \right) \\
   &= \left(\frac{f(x_n) - f(\alpha)}{f(x_{n-1}) - f(\alpha)} \right) \left(\frac{f'(\alpha) + (x_{n-1} - \alpha)f''(\xi_3)}{f'(\alpha) + (c_n - \alpha)f''(\xi_4)} \right) \\
   &= \left(\frac{\frac{f(x_n) - f(\alpha)}{x_n - \alpha}}{\frac{f(x_{n-1}) - f(\alpha)}{x_{n-1} - \alpha}} \right) \left(\frac{f'(\alpha) + (x_{n-1} - \alpha)f''(\xi_3)}{f'(\alpha) + (c_n - \alpha)f''(\xi_4)} \right) \\
   &= \left(\frac{f'(\eta_n)}{f'(\eta_{n-1})} \right) \left(\frac{f'(\alpha) + (x_{n-1} - \alpha)f''(\xi_3)}{f'(\alpha) + (c_n - \alpha)f''(\xi_4)} \right),
   \end{aligned}
   $$

   where η_n is between x_n and α, η_{n-1} is between x_{n-1} and α, ξ_3 is between x_{n-1} and α, and ξ_4 is between c_n and α. Therefore, as $n \to \infty$, we have $\eta_n \to \alpha$, $\eta_{n-1} \to \alpha$, $\xi_3 \to \alpha$ and $\xi_4 \to \alpha$. Thus,

   $$
   C_n \to \left(\frac{f'(\alpha)}{f'(\alpha)} \right) \left(\frac{f'(\alpha)}{f'(\alpha)} \right) = 1,
   $$

   so long as f'' is bounded.

2. We could also stop the iteration when $|f(x_n)|$ was sufficiently small. Use the Mean Value Theorem plus the fact that $f(\alpha) = 0$ to show that, if f' is continuous and non-zero near α, then there are constants c_1 and c_2 such that

   $$
   c_1 |f(x_n)| \leq |\alpha - x_n| \leq c_2 |f(x_n)|.
   $$

   Comment on this result.

3. Write a computer program that uses Newton's method to find the root of a given function, and apply this program to find the root of the following functions, using x_0 as given. Stop the iteration when the error as estimated by $|x_{n+1} - x_n|$ is less than 10^{-6}. Compare to your results for bisection.

   (a) $f(x) = 1 - 2xe^{-x/2}$, $x_0 = 0$;

   (b) $f(x) = 5 - x^{-1}$, $x_0 = \frac{1}{2}$;

   (c) $f(x) = x^3 - 2x - 5$, $x_0 = 2$;

   (d) $f(x) = e^x - 2$, $x_0 = 1$;

   (e) $f(x) = x - e^{-x}$, $x_0 = 1$;

   (f) $f(x) = x^6 - x - 1$, $x_0 = 1$;

   (g) $f(x) = x^2 - \sin x$, $x_0 = \frac{1}{2}$;

(h) $f(x) = x^3 - 2$, $x_0 = 1$;

(i) $f(x) = x + \tan x$, $x_0 = 3$;

(j) $f(x) = 2 - x^{-1} \ln x$, $x_0 = \frac{1}{3}$.

Solution: The results, obtained from a mid-range personal computer using MAT-LAB, are summarized for some functions in Table 3.4.

Table 3.4   Solutions to Problem 3.3.3.

$f(x)$	n	root
hline $x^3 - 2x - 5$	4	2.0946
$e^x - 2$	4	0.6931
$x^6 - x - 1$	5	1.1347
$x^2 - \sin x$	8	0.8767

4. Figure 3.2 shows the geometry of a planetary orbit[2] around the sun. The position of the sun is given by S, the position of the planet is given by P. Let x denote the angle defined by P_0OA, measured in radians. The dotted line is a circle concentric to the ellipse and having a radius equal to the major axis of the ellipse. Let T be the total orbital period of the planet, and let t be the time required for the planet to go from A to P. Then Kepler's equation from orbital mechanics, relating x and t, is

$$x - \epsilon \sin x = \frac{2\pi t}{T}.$$

Here ϵ is the eccentricity of the elliptical orbit (the extent to which it deviates from a circle). For an orbit of eccentricity $\epsilon = 0.01$ (roughly equivalent to that of the Earth), what is the value of x corresponding to $t = T/4$? What is the value of x corresponding to $t = T/8$? Use Newton's method to solve the required equation.

Solution: For $t = T/4$, Newton's method yielded $x = 1.580795827$ in 3 iterations; for $t = T/8$, we get $x = 0.7925194063$, also in 3 iterations. Both computations used $x_0 = 0$.

5. Consider now a highly eccentric orbit, such as that of a comet, for which $\epsilon = 0.9$ might be appropriate. What is the value of x corresponding to $t = T/4$? What is the value of x corresponding to $t = T/8$?

6. Consider the problem of putting water pipes far enough underground to avoid frozen pipes should the external temperature suddenly drop. Let T_0 be the initial temperature of the ground, and assume that the external air temperature suddenly drops to a new value $T < T_0$. Then a simple model of how the underground temperature responds to the change in external temperature tells us that

$$\frac{u(x,t) - T}{T_0 - T} = \operatorname{erf}\left(\frac{x}{2\sqrt{at}}\right).$$

Here $u(x,t)$ is the temperature at a depth x feet and time t seconds after the temperature change, and a is the thermal conductivity of the soil. Suppose that

---

[2]For the background material for this and the next problem, the author is indebted to the interesting calculus text: Alexander J. Hahn, *Basic Calculus; From Archimedes to Newton to its Role in Science*, Springer-Verlag, 2008.

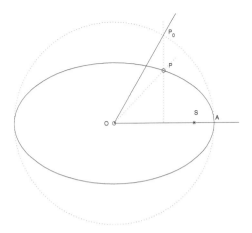

Figure 3.2   Orbital geometry for Problems 4 and 5

$a = 1.25 \times 10^{-6} \text{ft.}^2/\text{sec}$ and that the initial ground temperature is $T_0 = 40°$ (F). How deep must the pipe be buried to guarantee that the temperature does not reach 32° (F) for 30 days after a temperature shift to $T = 0°$ F? If your computing environment does not have an intrinsic error function, use the approximation presented in Problem 8 of §3.1.

7. In the previous problem, produce a plot of temperature u at depths of 6 feet and 12 feet, as a function of time (days), in response to a temperature shift of 40° for the following initial temperatures (in degrees Fahrenheit):

   (a) $T_0 = 40°$;

   (b) $T_0 = 50°$;

   (c) $T_0 = 60°$.

   Solution: This really is not a root-finding problem. Simply plot

   $$U(t) = u(6, t) = 40 + (T_0 - 40)\text{erf}\left(\frac{6}{2\sqrt{at}}\right)$$

   and similarly for $u(12, t)$, over the interval $0 \le t \le 86,400$. This produces a plot in which time is measured in seconds; it is an easy change of scale to convert it to days.

8. Use Newton's method to solve Problem 6 from §3.1.

9. Repeat the above, assuming that the mortgage is only 15 years in length.

   Solution: This amounts to a root finding problem for

   $$f(r) = L - \frac{12M}{r}[1 - (1 + (r/12))^{-12m}]$$

   for $L = 150$, $M = 0.6$, and $m = 15$. Newton's method finds a negative value of r, which implies that there is no interest rate which meets the specified conditions.

$$\triangleleft \bullet \bullet \triangleright$$

3.4 APPLICATION: DIVISION USING NEWTON'S METHOD

Exercises:

1. Test the method derived in this section by using it to approximate $1/0.75 = 1.333\ldots$, with x_0 as suggested in the text. Don't write a computer program, just use a hand calculator.

    Solution: We get the following sequence of values:

    $$x_1 = 1.312500,$$
    $$x_2 = 1.333007812,$$
    $$x_3 = 1.333333254,$$
    $$x_4 = 1.333333333,$$
    $$x_5 = 1.333333333.$$

2. Repeat the above for $a = 2/3$.

3. Repeat the above for $a = 0.6$.

    Solution: This time we get:

    $$x_1 = 1.656,$$
    $$x_2 = 1.6665984,$$
    $$x_3 = 1.666666664,$$
    $$x_4 = 1.666666667,$$
    $$x_5 = 1.666666667.$$

4. Based on the material in this section, write each of the following numbers in the form

    $$a = b \times 2^k,$$

    where $b \in [\frac{1}{2}, 1]$, and then use Newton's method to find the reciprocal of b and hence of a.

    (a) $a = 7$;

    Solution: We have
    $$a = (7/8) \times 2^3,$$
    so $b = 7/8$. Newton's method produces $b^{-1} = 1.142857143$ in four iterations, therefore $a^{-1} = b^{-1} \times 0.125 = 0.1428571429$.

    (b) $a = \pi$;

    (c) $a = 6$;

    (d) $a = 5$;

Solution: We have

$$5 = (5/2^3) \times 2^3 = 5/8 \times 8,$$

so $b = 5/8 = 0.625$. From the text, we take

$$x_0 = 3 - 2(0.625) = 1.75.$$

We get the results in Table 3.5. Notice that we converge much faster than expected, and that we have

$$1/5 = 1/1.6 \times 8 = 0.2,$$

which is the correct result.

Table 3.5   Newton iteration for Problem 4d, §3.4.

n	x_n
1	1.58594
2	1.59988
3	1.60000
4	1.60000
5	1.60000
6	1.60000

(e) $a = 3$.

Solution: We have

$$a = (3/4) \times 2^2,$$

so $b = 3/4$. Newton's method produces $b^{-1} = 1.333333333$ in four iterations, therefore $a^{-1} = b^{-1} \times 0.25 = 0.333333333$.

Be sure to use the initial value as generated in this section and only do as many iterations as are necessary for 10^{-16} accuracy. Compare your values to the intrinsic reciprocal function on your computer or calculator. Note this can be done on computer or calculator app.

5. Test the method derived in this section by writing a computer program to implement it. Test the program by having it compute $1/0.75 = 1.333...$, with x_0 as suggested in the text, as well as $1/(2/3) = 1.5$.

Solution: The following is a MATLAB script that accomplishes the desired task.

```
function r = recip(a)
    x = 3 - 2*a;
    for k=1:6
        x = x*(2 - a*x);
    end
    r = x;
```

6. How close an initial guess is needed for the method to converge to within a relative error of 10^{-14} in only *three* iterations?

Solution: To get the relative error after 3 iterations less than 10^{-14}, we need to have

$$\left|\frac{e_3}{\alpha}\right| = \left(\frac{e_0}{\alpha}\right)^8 \le 10^{-14}.$$

Therefore, the initial error $|e_0|$ must satisfy

$$|e_0| \le \alpha 10^{-14/8} = 10^{-14/8} = 0.0178.$$

7. Consider the quadratic polynomial

$$q_2(x) = 4.328427157 - 6.058874583x + 2.745166048x^2.$$

What is the error when we use this to generate the initial guess? How many steps are required to get 10^{-20} accuracy? How many operations?

8. Repeat Problem 7, this time with the polynomial

$$q_3(x) = 5.742640783 - 12.11948252x + 11.14228488x^2 - 3.767968453x^3.$$

Solution: This time we are looking at the difference

$$g(x) = \frac{1}{x} - q_3(x) = \frac{1}{x} - 5.742640783 + 12.11948252x - 11.14228488x^2 + 3.767968453x^3.$$

The derivative is therefore

$$g'(x) = (-1/x^2) + 12.11948252 - 22.28456976x + 11.30390x^2.$$

Setting this equal to zero and solving leads us to the polynomial equation

$$-1 + 12.11948252x^2 - 22.28456976x^3 + 11.30390x^4 = 0.$$

Plotting this shows there are three roots on the interval $[1/2, 1]$: $\alpha_1 \approx 0.56$, $\alpha_2 \approx 0.73$, and $\alpha_3 \approx 0.92$. Using Newton's method to refine each of these values gives us

$$\alpha_1 = 0.5621750583, \quad \alpha_2 = 0.7285533588, \quad \alpha_3 = .9163783254.$$

These are our critical points, at which we need to evaluate the error $g(x)$ (along with the endpoints). We get

$$g(1/2) = 0.0025253136$$
$$g(1) = 0.002525310$$
$$g(\alpha_1) = -0.0025253184$$
$$g(\alpha_2) = 0.002525312$$
$$g(\alpha_3) = -0.002525321,$$

thus showing that the maximum value of $|g(x)|$ — hence, the maximum value of the initial error using q_3 to compute x_0 — is about 0.0025. To get 10^{-20} accuracy now requires only 3 iterations, and a total of 15 operations, again.

9. Modify your program from Problem 5 to use each of p_2, q_2, and q_3 to compute $1/a$ for $a = 0.75$ and $a = 0.6$. Compare to the values produced by ordinary division on your computer. (Remember to use Horner's rule to evaluate the polynomials!)

10. Modify the numerical method to directly compute the ratio b/a, rather than just the reciprocal of a. Is the error analysis affected? (Note: Your algorithm should approximate the ratio b/a without any divisions at all.)

11. Test your modification by using it to compute the value of the ratio $4/5 = 0.8$.

3.5   THE NEWTON ERROR FORMULA

Exercises:

1. If f is such that $|f''(x)| \leq 3$ for all x and $|f'(x)| \geq 1$ for all x, and if the initial error in Newton's method is less than $\frac{1}{2}$, what is an upper bound on the error at each of the first three steps? Hint: Use the Newton error formula from this section.

   Solution: We have, from the Newton error formula,

   $$\alpha - x_{n+1} = -\frac{f''(\xi_n)}{2f'(x_n)}$$

   so that, using $e_n = |\alpha - x_n|$,

   $$e_1 \leq e_0^2 \frac{3}{2 \times 1} \leq (1/4)(3/2) = 3/8;$$

   $$e_2 \leq e_1^2 \frac{3}{2 \times 1} \leq (9/64)(3/2) = 27/128;$$

   $$e_3 \leq e_2^2 \frac{3}{2 \times 1} \leq (729/16384)(3/2) = 2187/32768 = 0.0667\ldots \ .$$

2. If f is now such that $|f''(x)| \leq 4$ for all x but $|f'(x)| \geq 2$ for all x, and if the initial error in Newton's method is less than $\frac{1}{3}$, what is an upper bound on the error at each of the first three steps?

3. Consider the left-hand column of data in Table 3.6 (Table 3.4 in the text.). Supposedly this comes from applying Newton's method to a smooth function whose derivative does not vanish at the root. Use the limit result (3.14) to determine whether or not the program is working. Hint: Use $\alpha \approx x_7$.

   Solution: Use $\alpha \approx x_7$ from the table, and compute values of the ratio

   $$r_n = \frac{\alpha - x_{n+1}}{(\alpha - x_n)^2}.$$

   If the method is converging according to the theory, this ratio should approach a constant value. For the data in this table we get

   $$\begin{aligned} r_1 &= -0.09523809523810, & r_2 &= -0.16122840690979, \\ r_3 &= -0.21215771728295, & r_4 &= -0.22329830363589, \\ r_5 &= -0.22031852835393. \end{aligned}$$

Table 3.6 Data for Problems 3 and 4.

n	x_n (Problem 3)	x_n (Problem 4)
0	10.0	0.000000000000
1	5.25	0.626665778573
2	3.1011904761905	0.889216318667
3	2.3567372726442	0.970768039664
4	2.2391572227372	0.992585155212
5	2.2360701085329	0.998139415613
6	2.2360679775008	0.999534421170
7	2.2360679774998	0.999883578197
8	N/A	0.999970892852
9	N/A	0.999992723105
10	N/A	0.999998180783

(Because our approximate α is x_7 we can't use r_6.) Since these values are settling down around -0.22, we are confidant that the program is working properly.

4. Repeat the previous exercise, using the right-hand column of data in Table 3.6.

5. Apply Newton's method to find the root of $f(x) = 2 - e^x$, for which the exact solution is $\alpha = \ln 2$. Perform the following experiments:

   (a) Compute the ratio

   $$R_n = \frac{\alpha - x_{n+1}}{(\alpha - x_n)^2}$$

   and observe whether or not it converges to the correct value as $n \to \infty$.

   (b) Compute the modified ratio

   $$R_n(p) = \frac{\alpha - x_{n+1}}{(\alpha - x_n)^p}$$

   for various $p \neq 2$, but near 2. What happens? Comment, in the light of the definition of order of convergence.

   Solution: For $p = 2$, the ratio will converge to the theoretical value of $r_\infty = -f''(\alpha)/2f'(\alpha) = -0.5$. For $p > 2$, the denominator goes to zero too fast so the ratio will "blow up." For $p < 2$, the denominator will not go to zero fast enough, so the ratio will go to zero.

6. Consider applying Newton's method to find the root of the function $f(x) = 4x - \cos x$. Assume we want accuracy to within 10^{-8}. Use the Newton error estimate (3.12) to show that the iteration will converge for all $x_0 \in [-2, 2]$. How many iterations will be needed for the iteration to converge? Compare with the corresponding results for bisection, using an initial interval of $[-2, 2]$.

7. Verify by actual computation that your results in the previous problem were correct, i.e., apply Newton's method to $f(x) = 4x - \cos x$; do you converge to the specified

accuracy in the correct number of iterations, and does this convergence appear to be occurring for all choices of x_0?

Solution: What should be done here is to compute the root of f for a variety of x_0 values on the interval $[-2, 2]$ to verify that the convergence does occur as fast as predicted for *all* $x_0 \in [-2, 2]$.

8. Consider now the function $f(x) = 7x - \cos(2\pi x)$. Show that a root exists on the interval $[0, 1]$, and then use the Newton error estimate to determine how close x_0 has to be to the root to guarantee convergence.

9. Investigate your results in the previous problem by applying Newton's method to $f(x) = 7x - \cos(2\pi x)$, using several choices of x_0 within the interval $[0, 1]$. Comment on your results in light of the theory of the method. Note: This can be done by a very modest modification of your existing Newton program.

Solution: We first observe that $f(0) = -1$ and $f(1) = 6$, so a root exists on $[0, 1]$. To test how close we have to be for convergence, I first used my bisection code to find this root. I then modified my Newton program to cycle through different values of x_0, printing out the error $\alpha - x_1$ as a fraction of $\alpha - x_0$, and also recording whether or not the iteration ultimately converged. My output is given in Table 3.7. Note that if the initial guess was far from the actual root of 0.110047, then convergence was certainly delayed. In particular, for $x_0 \in [0.5, 0.9]$, x_1 was not in $[0, 1]$. While the iteration did eventually converge for most choices of x_0, for $x_0 = 0.6$ and $x_0 = 0.8$ the iteration has not yet converged after 20 iterations and I would hazard the guess that it is unlikely to converge, ever.

Table 3.7 Newton iteration for $f(x) = 7x - \cos 2\pi x$.

x_0	x_1	x_5	x_{20}
0.000000	0.142857	0.110047	0.110047
0.100000	0.110195	0.110047	0.110047
0.200000	0.115921	0.110047	0.110047
0.300000	0.114343	0.110047	0.110047
0.400000	0.624931×10^{-1}	0.110047	0.110047
0.500000	-0.142857	-0.290451	0.110047
0.600000	-0.914746	-0.145064	-1.59073
0.700000	-4.38526	0.201481	0.110047
0.800000	-4.36528	0.378386×10^2	0.544820×10^3
0.900000	-0.760494	-0.262859	0.110047
10.00000	0.142857	0.110047	0.110047

10. Show that if x_n is a sequence converging linearly to a value α, then the constant C in (3.16) must satisfy $|C| < 1$. Hint: Assume $|C| > 1$ and prove a contradiction to the convergence assumption.

Solution: If $|C| > 1$ then, for all n sufficiently large, we have $|\alpha - x_{n+1}| > |\alpha - x_n|$, which means convergence cannot occur.

11. Explain, in your own words, why the assumptions $f'(x_n) \approx f'(\alpha)$ and $f''(\xi_n) \approx f''(\alpha)$ are valid if we have $x_n \to \alpha$.

Solution: If f' and f'' are continuous, then we have that

$$\lim_{n\to\infty} f(x_n) = f\left(\lim_{n\to\infty} x_n\right)$$

and similarly for f''.

$$\lhd \bullet \bullet \bullet \rhd$$

3.6 NEWTON'S METHOD: THEORY AND CONVERGENCE

Exercises:

1. Consider a function f which satisfies the properties:

   (a) There exists a unique root $\alpha \in [0, 1]$;
   (b) For all real x we have $f'(x) \geq 2$ and $0 \leq f''(x) \leq 3$.

   Show that for $x_0 = \frac{1}{2}$, Newton's Method will converge to within 10^{-6} of the actual root in four iterations. How long would bisection take to achieve this accuracy?

   Solution: The Newton error formula, together with the given bounds on the derivatives, tells us that

   $$|\alpha - x_{n+1}| \leq \left(\frac{3}{4}\right)|\alpha - x_n|^2.$$

   Therefore, for $x_0 = \frac{1}{2}$, we have $|\alpha - x_0| \leq \frac{1}{2}$ and hence,

   $$|\alpha - x_1| \leq (3/4)(1/2)^2 = 3/16;$$

   $$|\alpha - x_2| \leq (3/4)(3/16)^2 = 27/1024;$$

   $$|\alpha - x_3| \leq (3/4)(27/1024)^2 = 0.000521421;$$

   $$|\alpha - x_4| \leq (3/4)(0.000521421)^2 = 0.204 \times 10^{-6}.$$

   Bisection would take 20 iterations to obtain this accuracy, based on the formula (3.2).

2. Consider a function f which satisfies the properties:

   (a) There exists a unique root $\alpha \in [2, 3]$;
   (b) For all real x we have $f'(x) \geq 3$ and $0 \leq f''(x) \leq 5$.

   Using $x_0 = 5/2$, will Newton's method converge, and if so, how many iterations are required to get 10^{-4} accuracy?

   Solution: We have

   $$|\alpha - x_{n+1}| \leq \left(\frac{5}{6}\right)|\alpha - x_n|^2.$$

   For $x_0 = 5/2$, we have $|\alpha - x_0| \leq \frac{1}{2}$ and hence,

   $$|\alpha - x_1| \leq (5/6)(1/2)^2 = 5/24;$$

$$|\alpha - x_2| \le (5/6)(5/24)^2 = 0.3616898148 \times 10^{-1};$$
$$|\alpha - x_3| \le (5/6)(.3616898148e - 1)^2 = 0.1090162684 \times 10^{-2};$$
$$|\alpha - x_4| \le (5/6)(.1090162684e - 2)^2 = 0.9903788983 \times 10^{-6}.$$

So the iteration does converge, and four iterations are enough to obtain the desired accuracy.

3. Repeat the above, this time aiming for 10^{-20} accuracy.

Solution: Continuing the calculation in the above problem, we see that

$$|\alpha - x_6| \le 7.2936 \times 10^{-28}$$

so that the sixth iterate will have error less than 10^{-20}.

4. Consider a function f which satisfies the properties:

   (a) There exists a unique root $\alpha \in [-1, 3]$;
   (b) For all real x we have $f'(x) \ge 4$ and $-6 \le f''(x) \le 3$.

   Using $x_0 = 1$, will Newton's method converge, and if so, how many iterations are required to get 10^{-4} accuracy?

   Solution: The Newton error formula gives us

$$|\alpha - x_{n+1}| \le \frac{6}{2 \times 4}(\alpha - x_n)^2 = \frac{3}{4}(\alpha - x_n)^2.$$

   Since the initial error is bounded above by 2, we have

$$|\alpha - x_1| \le \frac{3}{4} \times 4 = 3 \ge |\alpha - x_0|,$$

   therefore we do not know if convergence follows.

5. Repeat the above, this time aiming for 10^{-20} accuracy.

   Solution: There's nothing to repeat, since we don't know if the iteration converges or not.

6. Consider a function that satisfies the following properties:

   (a) f is defined and twice continuously differentiable for all x;
   (b) f has a unique root $\alpha \in [-1, 1]$;
   (c) $|f'(x)| \ge 2$ for all x;
   (d) $|f''(x)| \le 5$ for all x.

   Can we conclude that Newton's method will converge for all $x_0 \in [-1, 1]$? If so, how many iterations are required to get 10^{-6} accuracy? If not, how many steps of bisection must we do to get the initial interval small enough so that Newton's method will converge? (For *any* choice of x_0.) How many total function evaluations (bisection plus Newton) are required to get 10^{-6} accuracy?

   Solution: We have $e_0 = |\alpha - x_0| \le 2$ and

$$e_{n+1} \le (1/2)e_n^2(5/2) = (5/4)e_n^2.$$

Therefore,
$$e_1 \leq (5/4)(2^2) = 5 > e_0,$$

so we cannot conclude that convergence will occur. If we do a single step of bisection, the initial Newton error will now be 1, and we will have
$$e_1 \leq (5/4)(1^2) = 5/4 > e_0,$$

so we need a second bisection step to get the e_0 for Newton less than $1/2$. Now we have
$$e_1 \leq (5/4)(1/4) = 5/16$$
$$e_2 \leq (5/4)(5/16)^2 = 125/1024$$

and so on. We get 10^{-6} accuracy after 5 iterations. Thus, it takes a total of 12 function evaluations.

7. Repeat the above for a function satisfying the properties:

   (a) f is defined and twice continuously differentiable for all x;

   (b) f has a unique root $\alpha \in [-1, 2]$;

   (c) $f'(x) \leq -3$ for all x;

   (d) $|f''(x)| \leq 4$ for all x.

   Solution: We have $e_0 \leq 3$ and
   $$e_{n+1} \leq \frac{1}{2} e_n^2 (4/3) = \frac{2}{3} e_n^2.$$

   Thus,
   $$e_1 \leq \frac{2}{3}(3^2) = 6 > e_0.$$

   After a single step of bisection, we have $e_0 \leq 3/2$ and
   $$e_1 \leq \frac{2}{3}(9/4) = 3/2 \geq e_0.$$

   So we need to take a second bisection step, in which case we have $e_0 \leq 3/4$ and
   $$e_1 \leq \frac{2}{3}(9/16) = 3/8.$$

   We converge to 10^{-6} accuracy in 5 Newton iterations, again using a total of 12 function evaluations.

8. Consider the function $f(x) = x - a \sin x - b$, where a and b are positive parameters, with $a < 1$. Will the initial guess $x_0 = b$ always lead to convergence? If not, what additional condition on a or the initial guess needs to be made?

   Solution: The Newton error formula, applied to this function, gives us
   $$|\alpha - x_{n+1}| = \frac{1}{2}(\alpha - x_n)^2 \left| \frac{a \sin \xi_n}{1 - a \cos x_n} \right|.$$

Since $a < 1$ this becomes

$$|\alpha - x_{n+1}| \leq \frac{1}{2}(\alpha - x_n)^2 \frac{a}{1-a}.$$

Convergence is implied by

$$\frac{1}{2}|\alpha - x_0|\frac{a}{1-a} = \frac{1}{2}|\alpha - b|\frac{a}{1-a} \leq 1,$$

which obviously will not be satisfied under all circumstances.

9. Write a program using Newton's method to find the root of $f(x) = 1 - e^{-x^2}$; the exact value is $\alpha = 0$. Compute the ratio

$$R_n = \frac{\alpha - x_{n+1}}{(\alpha - x_n)^2}$$

as the iteration progresses and comment (carefully!) on your results, in the light of the material in this section.

Solution: The iteration will converge, but slowly, and the quadratic convergence will be lost (thus the value of R_n will not converge). This occurs because $f'(\alpha) = 0$, thus a major condition of our theory does not hold.

10. A *monotone* function is one whose derivative never changes sign; the function is either always increasing or always decreasing. Show that a monotone function can have at most one root.

Solution: Suppose a monotone function has two roots, α_1 and α_2. Then Rolle's Theorem implies that there is a point ξ between the roots where $f'(\xi) = 0$, and this violates the monotonicity requirement. Thus there cannot be more than one root.

11. If f is a smooth monotone function with a root $x = \alpha$, will Newton's method always converge to this root, for any choice of x_0? (Either provide a counter-example or a valid proof.)

Solution: The arctangent function discussed earlier in Chapter 3 is a counterexample. It is monotone, and there exist values of x_0 such that Newton's method will not converge to the root $\alpha = 0$.

3.7 APPLICATION: COMPUTATION OF THE SQUARE ROOT

Exercises:

1. Based on the material in this section, write each of the following numbers in the form

$$a = b \times 2^k,$$

where $b \in [\frac{1}{4}, 1]$ and k is even, and then use Newton's method to find the square root of b and hence of a:

(a) $a = \pi$;

(b) $a = 5$;

(c) $a = 7$;

(d) $a = 3$;

(e) $a = 6$.

Be sure to use the initial value as generated in this section and only do as many iterations as are necessary for 10^{-16} accuracy. Compare your values to the intrinsic square root function on your computer or calculator. Note this can be done on a computer or hand calculator.

Solution: For the single case of (c), we have

$$7 = \frac{7}{16} \times 2^4$$

so we will be finding the square root of $7/16 = 0.4375$. The initial guess is given by

$$x_0 = (2(7/16) + 1)/3 = 0.625.$$

Newton's method then gives us

$$x_1 = 0.5 \times (0.625 + 0.4375/0.625) = 0.6625;$$

$$x_2 = 0.5 \times (0.6625 + 0.4375/0.6625) = 0.66143867924528;$$

$$x_3 = 0.5 \times (0.66143867924528 + 0.4375/0.66143867924528) = 0.66143782776670;$$

$$x_4 = 0.5 \times (0.66143782776670 + 0.4375/0.66143782776670) = 0.66143782776615;$$

$$x_5 = 0.5 \times (0.66143782776615 + 0.4375/0.66143782776615) = 0.66143782776615.$$

Thus,

$$\sqrt{7} = \sqrt{7/16} \times 2^2 = 0.66143782776615 \times 4 = 2.64575131106459.$$

On my old calculator the value is $\sqrt{7} = 2.645751311$. MATLAB on my PC at home reports a value identical to what we got here.

2. In the example of this section, how many iterations are required for the *absolute* error to be less than 2^{-48}?

Solution: The absolute error satisfies

$$|\sqrt{b} - x_{n+1}| \le \frac{1}{2\sqrt{b}}|\sqrt{b} - x_n|^2 \le |\sqrt{b} - x_n|^2.$$

So it is easy to show that, with an initial error of no more than $9/64$, we will converge to the specified accuracy in 5 iterations.

3. How accurate is the result in only 3 iterations, using the initial guess used here?

    Solution: We have

    $$\left|\frac{\sqrt{b} - x_1}{\sqrt{b}}\right| \le \frac{1}{2}\left|\frac{9/64}{\sqrt{b}}\right|^2 \le \frac{1}{2}(9/32)^2 = 81/2048 \le 0.04.$$

    Thus

    $$\left|\frac{\sqrt{b} - x_2}{\sqrt{b}}\right| \le \frac{1}{2}(0.04)^2 = 0.0008$$

    and

    $$\left|\frac{\sqrt{b} - x_3}{\sqrt{b}}\right| \le \frac{1}{2}(0.0008)^2 = 0.32 \times 10^{-6}.$$

4. What does the initial error have to be for the relative error after only *two* iterations to be less than 10^{-14}?

    Solution: We require that

    $$2\left(\frac{\sqrt{b} - x_0}{\sqrt{b}}\right)^4 \le 10^{-14}.$$

    Thus the initial relative error has to be less than $(2 \times 10^{-14})^{1/4} = 0.0003761$. Since the relative error is bounded above by as much as twice the absolute error, we have to have $|\sqrt{b} - x_0| \le 0.0001880$.

5. Extend the derivation and analysis of this section to the problem of computing *cube* roots by solving for the roots of $f(x) = x^3 - a$. Be sure to cover all the major points that were covered in the text. Is using linear interpolation to get x_0 accurate enough to guarantee convergence of the iteration?

    Solution The iteration is easily found to be

    $$x_{n+1} = \frac{1}{3}\left(2x_n + \frac{a}{x_n^2}\right).$$

    where we now assume that $a \in [1/8, 1]$, perhaps by using some more exponent shifting. The Newton error formula gives us that

    $$\left|\frac{\alpha - x_{n+1}}{\alpha}\right| \le \frac{1}{2}\left|\frac{\alpha - x_n}{\alpha}\right|^2\left|\frac{6a\xi_n}{3x_n^2}\right| = \left|\frac{a\xi_n}{x_n^2}\right|\left|\frac{\alpha - x_n}{\alpha}\right|^2,$$

    where ξ_n is between x_n and $\alpha = a^{1/3}$. The same argument as used in the text shows that, for $n \ge 1$, $x_n \ge \alpha$, therefore we have

    $$\left|\frac{\alpha - x_{n+1}}{\alpha}\right| \le \left|\frac{\alpha - x_n}{\alpha}\right|^2.$$

The initial guess polynomial is the linear interpolate connecting $(1/8, 1/2)$ and $(1, 1)$, and works out to be

$$p_1(a) = (4a + 3)/7.$$

The error in this approximation can be bounded according to

$$|x_0 - p_1(a)| \le (1/8)(7/8)^2 \max_{t \in [1/8, 1]} |(2/9)t^{-5/3}| \le (1/8)(49/64)(2/9)(8)^{5/3} = 0.681.$$

Using this as an upper bound on the initial error, we find that

$$\left| \frac{\alpha - x_1}{\alpha} \right| \le (0.681)^2 = 0.4631558643,$$

and we get convergence (to 10^{-14} accuracy) in 7 iterations.

The error estimate for the initial guess is very conservative here. Looking at a plot of $x^{1/3} - p_1(x)$ shows that the largest error is on the order of 0.09; using this in our bound leads to convergence in 4 iterations.

6. Consider using the polynomial

$$p_2(x) = \frac{9}{35} + \frac{22}{21}x - \frac{32}{105}x^2$$

to generate the initial guess. What is the maximum error in $|\sqrt{x} - p_2(x)|$ over the interval $[\frac{1}{4}, 1]$? How many iterations (and hence, how many operations) are needed to get accuracy of 10^{-16}? (Feel free to compute the maximum value experimentally, but you must justify the accuracy of your value in this case with some kind of argument.)

Solution: Experimentally, the error $|\sqrt{x} - p_2(x)|$ is bounded above by about 0.006. This can be confirmed using ordinary calculus by finding the extreme values of the function $g(x) = \sqrt{x} - p_2(x)$. With this as the initial absolute error, we have that the relative error behaves as follows:

$$\left| \frac{\sqrt{b} - x_1}{\sqrt{b}} \right| \le \frac{1}{2} \left| \frac{0.006}{\sqrt{b}} \right|^2 \le \frac{1}{2}(0.012)^2 = 0.0000720,$$

$$\left| \frac{\sqrt{b} - x_2}{\sqrt{b}} \right| \le 2.6 \times 10^{-8},$$

$$\left| \frac{\sqrt{b} - x_3}{\sqrt{b}} \right| \le 0.336 \times 10^{-17},$$

$$\left| \frac{\sqrt{b} - x_4}{\sqrt{b}} \right| \le 0.564 \times 10^{-35}.$$

The total number of operations to get this accuracy is 4 to get the initial guess plus 3 in each of 4 iterations, for a total of 16 operations.

7. Repeat the above using

$$q_2(x) = 0.2645796916 + 1.0302824818x - 0.2983646911x^2.$$

Solution: The empirical error is about 0.0036, leading to convergence within 10^{-16} in 4 iterations, for a total of 16 operations, again.

8. Repeat again, using

$$q_1(x) = 0.647941993t + 0.3667102689.$$

Solution: The empirical error is about 0.03, leading to convergence within 10^{-16} in 4 iterations, for a total of 15 operations, because computing the initial guess is cheaper this time.

9. Re-implement your program using p_2, q_2, and q_1 to generate the initial guess, applying it to each of the values in Problem 1. Compare your results to the square root function on your system.

10. If we use a piecewise linear interpolation to construct x_0, using the nodes $\frac{1}{4}$, $\frac{9}{16}$, and 1, what is the initial error and how many iterations are now needed to get 10^{-16} accuracy? How many operations are involved in this computation?

Solution: This is, to great extent, the same as Exercise 10 of §2.4. The maximum initial error is now 0.0244, which leads to convergence in 4 iterations, and a total of 15 operations (plus one comparison to decide which of the two polynomial pieces to use).

◁ • • • ▷

3.8 THE SECANT METHOD: DERIVATION AND EXAMPLES

Exercises:

1. Do three steps of the secant method for $f(x) = x^3 - 2$, using $x_0 = 0$, $x_1 = 1$.
   Solution: We get $x_2 = 2$, $x_3 = 1.142857143$, and $x_4 = 1.209677419$.

2. Repeat the above using $x_0 = 1$, $x_1 = 0$. Comment.
   Solution: We get $x_2 = 2$, $x_3 = 0.5$, $x_4 = 0.8571$; note that simply switching x_0 and x_1 resulted in substantially different results.

3. Apply the secant method to the same functions as in Problem 3 of §3.1, using x_0, x_1 equal to the endpoints of the given interval. Stop the iteration when the error as estimated by $|x_n - x_{n-1}|$ is less than 10^{-6}. Compare to your results for Newton and bisection in the previous problems.

   Solution: For the particular case of (d), Table 3.8 shows the results of my program; note that we find a different root than Newton or bisection found. This can happen with root-finding methods. If we reverse x_1 and x_0, then we find the same root as the other methods did.

Table 3.8 Secant iteration for $f(x) = x^6 - x - 1$.

n	x_n	$|x_n - x_{n-1}|$
0	0.000000000000	N/A
1	2.000000000000	2.000000000000
2	0.032258064516	0.032258064516
3	-1.000000034929	1.032258099446
4	-0.492063446441	0.507936588488
5	-0.659956919149	0.167893472708
6	-0.842841677613	0.182884758464
7	-0.762579762950	0.080261914663
8	-0.776093797503	0.013514034553
9	-0.778152697560	0.002058900057
10	-0.778089343144	0.000063354416
11	-0.778089598646	0.000000255502

4. For the secant method, prove that

$$\alpha - x_{n+1} = C_n(x_{n+1} - x_n),$$

where $C_n \to 1$ as $n \to \infty$, so long as the iteration converges. (Hint: Follow what we did in §3.3 for Newton's Method.)

Solution: We have

$$
\begin{aligned}
\alpha - x_n &= -\frac{f(x_n)}{f'(c_n)}, \\
&= -\frac{f(x_n)}{f'(c_n)} \frac{f(x_n) - f(x_{n-1})}{x_n - x_{n-1}} \frac{x_n - x_{n-1}}{f(x_n) - f(x_{n-1})} \\
&= -\frac{f(x_n)f'(\xi_n)}{f(x_n)f'(c_n)}(x_{n+1} - x_n) \\
&= C_n(x_{n+1} - x_n),
\end{aligned}
$$

where

$$C_n = -\frac{f'(\xi_n)}{f'(c_n)},$$

for ξ_n between x_n and x_{n-1} and c_n between x_n and α.

5. Assume (3.29) and prove that if the secant method converges, then it is superlinear.

Solution: We have

$$\alpha - x_{n+1} = -\frac{1}{2}(\alpha - x_n)(\alpha - x_{n-1})\frac{f''(\xi_n)}{f'(\eta_n)}$$

so that

$$\frac{\alpha - x_{n+1}}{(\alpha - x_n)} = -\frac{1}{2}(\alpha - x_{n-1})\frac{f''(\xi_n)}{f'(\eta_n)}.$$

As $x_n \to \alpha$ we also have $x_{n-1} \to \alpha$ and $\eta_n, \xi_n \to \alpha$. Therefore,

$$\frac{\alpha - x_{n+1}}{(\alpha - x_n)} \to 0,$$

which means the method is superlinear.

6. Assume (3.29) and consider a function f such that

    (a) There is a unique root on the interval $[0, 4]$;

    (b) $|f''(x)| \leq 2$ for all $x \in [0, 4]$;

    (c) $f'(x) \geq 5$ for all $x \in [0, 4]$.

Can we prove that the secant iteration will converge for any $x_0, x_1 \in [0, 4]$? If so, how many iterations are required to get an error that is less than 10^{-8}? If convergence is not guaranteed for all $x \in [0, 4]$, how many steps of bisection are needed before convergence *will* be guaranteed for secant?

Solution: We have

$$e_{n+1} \leq \frac{1}{2} e_n e_{n-1}(2/5).$$

Since the initial errors are no worse than 4, then, we have

$$e_2 \leq \frac{1}{2}(16)(2/5) = 16/5,$$

$$e_3 \leq \frac{1}{2}(16/5)(4)(2/5) = 16/25,$$

and we converge to the desired accuracy in 10 iterations.

7. Repeat the above problem under the assumptions:

    (a) There is a unique root on the interval $[0, 9]$;

    (b) $|f''(x)| \leq 6$ for all $x \in [0, 9]$;

    (c) $f'(x) \geq 2$ for all $x \in 0, 9]$.

Can we prove that the secant iteration will converge for any $x_0, x_1 \in [0, 9]$? If so, how many iterations are required to get an error that is less than 10^{-8}? If convergence is not guaranteed for all $x \in [0, 9]$, how many steps of bisection are needed before convergence *will* be guaranteed for secant?

Solution: We have

$$e_{n+1} \leq \frac{1}{2} e_n e_{n-1}(2/5).$$

Since the initial errors are no worse than 9, then, we have

$$e_2 \leq \frac{1}{2}(81)(3) = 243/2;$$

so we will not converge. Doing one step of bisection reduces the original secant interval to length 4.5, so we then have

$$e_2 \leq \frac{1}{2}(9/2)^2(3) = 243/8,$$

which is still too large. Another step of bisection give us

$$e_2 \leq \frac{1}{2}(9/4)^2(3) = 243/32,$$

and a third bisection yields

$$e_2 \leq \frac{1}{2}(9/8)^2(3) = 243/128,$$

still not small enough. Finally, a fourth bisection step gives an initial interval of length $9/16$ and a first step error of

$$e_2 \leq \frac{1}{2}(9/16)^2(3) = 243/512.$$

Convergence then occurs in 10 secant iterations.

8. Repeat Problem 8 of §3.1, except this time find α for the set of θ values defined by

$$\theta_k = 10^{k/4},$$

for k ranging from -24 all the way to 24. Construct a plot of α vs. $\log_{10} \theta$.

Solution: For $k = -24, \alpha = 0.00094129631308$, for $k = -23, \alpha = 0.00125518056472$, for $k = 1, \alpha = 0.92775227544590$, for $k = 2, \alpha = 1.09242933874440$. The plot is given in Figure 3.3.

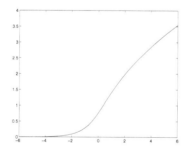

Figure 3.3   Solution for Problem 3.8.10.

9. Comment, in your own words, on the differences between the secant method and regula-falsi (see Problem 9 of §3.1).

Solution: Regula-falsi chooses the two approximate values in order to guarantee that the root is "bracketed," whereas the secant method chooses the most accurate (i.e., most recent) approximations, regardless of whether or not they bracket the root.

◁ • • • ▷

3.9 FIXED POINT ITERATION

Exercises:

1. Do three steps of each of the following fixed point iterations, using the indicated x_0.

   (a) $x_{n+1} = \cos x_n$, $x_0 = 0$ (be sure to set your calculator in radians);
      Solution: $x_1 = 1$, $x_2 = 0.5403023059$, $x_3 = 0.8575532158$

   (b) $x_{n+1} = e^{-x_n}$, $x_0 = 0$;

   (c) $x_{n+1} = \ln(1 + x_n)$, $x_0 = 1/2$;
      Solution: $x_1 = 0.4054651081$, $x_2 = 0.3403682857$, $x_3 = 0.2929444165$

   (d) $x_{n+1} = \frac{1}{2}(x_n + 3/x_n)$, $x_0 = 3$.

2. Let $Y = 1/2$ be fixed, and take $h = \frac{1}{8}$. Do three steps of the following fixed point iteration

$$y_{n+1} = Y + \frac{1}{2}h\left(-Y \ln Y - y_n \ln y_n\right)$$

   using $y_0 = Y$.

   Solution: $y_1 = 0.5433216988$, $y_2 = 0.5423768123$, $y_3 = 0.5423997893$.

3. Let $Y_0 = 1/2$ and $Y_1 = 0.54332169878500$ be fixed, and take $h = \frac{1}{8}$. Do three steps of the fixed point iteration,

$$y_{n+1} = \frac{4}{3}Y_1 - \frac{1}{3}Y_0 - 2hy_n \ln y_n$$

   using $y_0 = Y_1$.

   Solution: $y_1 = 0.6406261163$, $y_2 = 0.6290814526$, $y_3 = 0.6306562188$.

4. Consider the fixed-point iteration $x_{n+1} = 1 + e^{-x_n}$. Show that this iteration converges for any $x_0 \in [1, 2]$. How many iterations does the theory predict it will take to achieve 10^{-5} accuracy?

   Solution: We have $g(x) = 1 + e^{-x}$, and therefore $1 \le g(x) \le 2$ for all $x \in [1, 2]$. In addition, $g'(x) = -e^{-x}$ so $|g(x)| \le e^{-1}$ for all $x \in [1, 2]$. We can now apply Theorem 3.5 to get that a unique fixed point exists, and the iteration $x_{n+1} = g(x_n)$ converges, with the error satisfying

$$|\alpha - x_n| \le \frac{e^{-n}}{1 - e^{-1}}|x_1 - x_0| \le 1.6 \times e^{-n}.$$

   It therefore takes $n = 12$ iterations to converge to the specified accuracy.

5. For each function listed below, find an interval $[a, b]$ such that $g([a, b]) \subset [a, b]$. Draw a graph of $y = g(x)$ and $y = x$ over this interval, and confirm that a fixed point exists there. Estimate (by eye) the value of the fixed point, and use this as a starting value for a fixed point iteration. Does the iteration converge? Explain.

   (a) $g(x) = \frac{1}{2}(x + \frac{2}{x})$;
      Solution: $[a, b] = [1, 3/2]$.

(b) $g(x) = x + e^{-x} - \frac{1}{4}$;

(c) $g(x) = \cos x$;

   Solution: $[a, b] = [0, 1]$.

(d) $g(x) = 1 + e^{-x}$;

   Solution: $[a, b] = [1, 2]$.

(e) $g(x) = \frac{1}{2}(x^2 + 1)$.

6. Let $h(x) = 1 - x^2/4$. Show that this function has a root at $x = \alpha = 2$. Now, using $x_0 = 1/2$, do the iteration $x_{n+1} = h(x_n)$ to approximate the fixed point of h. Comment on your results.

   Solution: Since $1 - x^2/4 = 0 \Rightarrow x = 2$, h obviously has a root at $x = 2$. On the other hand, the iteration $x_{n+1} = h(x_n)$ converges to $\alpha = 0.8284271247$. The point is that α is a fixed point of h, not a root of h.

7. Let $h(x) = 3 - e^{-x}$. Using $x_0 = 2$, perform as many iterations of Newton's method as are needed to accurately approximate the root of h. Then, using the same initial point, do the iteration $x_{n+1} = h(x_n)$ to approximate the fixed point of h. Comment on your results.

   Solution: This is essentially making the same point as the previous problem. The root is -1.098612289 (approximately), but the fixed point is 2.947530903 (approximately).

8. Use fixed point iteration to find a value of $x \in [1, 2]$ such that $2 \sin \pi x + x = 0$.

9. For $a > 0$, consider the iteration defined by

$$x_{n+1} = \frac{x_n^3 + x_n^2 - x_n a + a}{x_n^2 + 2x_n - a}.$$

   (a) For $x_0 = 3/2$ experiment with this iteration for $a = 4$ and $a = 2$. Based on these results, speculate as to what this iteration does. Try to prove this, and use the theorems of this section to establish a convergence rate.

   (b) Now experiment with this iteration using $x_0 = 2$ and $a = 5$. Compare your results to Newton's method.

10. Consider the iteration

$$x_{n+1} = \frac{(N-1)x^{N+1} + ax}{Nx^N}.$$

   Assume that this converges for integer N and any $a > 0$. What does it converge to? Use the theorems of this section to determine a convergence rate. Experiment with this iteration when $N = 3$ and $a = 8$, using $x_0 = 3/2$.

11. Consider the iteration defined by

$$x_{n+1} = x_n - f(x_n)\left[\frac{f(x_n)}{f(x_n + f(x_n)) - f(x_n)}\right].$$

   This is also sometimes known as Steffenson's method. Show that it is (locally) quadratically convergent.

Solution: This is a somewhat tricky application of Theorem 3.7. We have

$$g(x) = x - f(x) \left[\frac{f(x)}{f(x + f(x)) - f(x)} \right],$$

so that

$$g'(x) = 1 - f'(x) \left[\frac{f(x)}{f(x + f(x)) - f(x)} \right] - f(x) \left[\frac{f(x)}{f(x + f(x)) - f(x)} \right]'.$$

Now, for $x = \alpha$ the last term vanishes (because $f(\alpha) = 0$), so we can ignore it. The second term becomes indeterminate (it naively evaluates to $0/0$) so we have to employ L'Hôpital's rule to get $g'(\alpha) = 0$. We then have to show that $g''(\alpha) \neq 0$.

12. Apply Steffenson's method to find the root of $f(x) = 2 - e^x$, using $x_0 = 0$. Compare your convergence results to those in the text for Newton and the secant method.

    Solution: The first 4 iterates are

    $$x_1 = 0.5819767070, \quad x_2 = 0.6876240766, \quad x_3 = 0.6931320121, \quad x_4 = 0.6931471806.$$

    Clearly the iteration is converging to the same root as for Newton and secant, and with comparable speed.

13. Apply Steffenson's method to $f(x) = x^2 - a$ for the computation of $\sqrt{a}$. For the following values of a, how does the performance compare to Newton's method?

    (a) $a = 3$;

    (b) $a = 2$;

        Solution: Using $x_0 = 2$, the first three Newton iterates are $x_1 = 1.500000000$, $x_2 = 1.416666667$, and $x_3 = 1.414215686$; the first three Steffenson iterates (using the same x_0) are $x_1 = 1.666666667$, $x_2 = 1.477477478$, and $x_3 = 1.419177338$.

    (c) $a = \pi$.

◁ • • ▷

3.10 ROOTS OF POLYNOMIALS (PART 1)

Exercises:

1. Use the Durand-Kerner algorithm to find all the roots of the polynomial

   $$p(x) = x^4 - 10x^3 + 35x^2 - 50x + 24.$$

   You should get $\zeta_{1,2,3,4} = (1, 2, 3, 4)$.

   Solution: The author's code (below) converged in 11 iterations. Here are the ζ values for the first three iterations, using the roots of unity as the initial guesses:

   $$\zeta_j^{(1)} = -10.0000 + 3.5000i, \quad -2.7694 - 4.0214i, \quad 0.5263 - 0.6208i, \quad 1.0000 - 0.0000i,$$

$$\zeta_j^{(2)} = 9.0724{+}11.2615i, \quad 0.9705{-}3.5225i, \quad 0.6181{-}0.2987i, \quad 1.0000{-}0.0000i,$$

$$\zeta_j^{(3)} = 6.1229{+}3.0728i, \quad 3.0670{-}2.0642i, \quad 1.1368{-}0.0148i, \quad 1.0000{+}0.0000i.$$

Using the roots of unity as initial guesses actually simplified the problem a bit, since the initial guess for ζ_4 was exact!

The following code is set up as a function which returns the converged values of the roots.

Algorithm 3.1 *Code for Durand-Kerner*

```
function r = DKroots(p)
np = length(p);
n = np-1;
q = p/p(1);
k = [1:n];
zeta = exp(2*pi*i*k/n);
for j=1:20
old = zeta;
for k=1:n
qp = polyval(q,zeta(k));
denom = 1.0;
for kk=1:n
if k ~= kk
denom = denom*(zeta(k) - zeta(kk));
end
end
zeta(k) = zeta(k) - qp/denom;
end
zeta    %Delete this line to avoid excess output
if abs(zeta - old) <= 1.e-14
break
end
end
r = zeta;
```

---

2. Use the Durand-Kerner algorithm to find all the roots of the following polynomials (Feel free to use MATLAB'S roots command to check your results):

   (a) $p(x) = x^6 + x^5 + x^4 + x^3 + x^2 + x + 1$;

   (b) $p(x) = x^6 - x^5 - 1$;

   (c) $p(x) = x^9 - x^8 - 1$;

   (d) $p(x) = x^5 - 1$.

Solution: For (a) we get $\zeta = 0.6235{+}0.7818i$, $\quad -0.2225{+}0.9749i$, $\quad -0.9010{+}0.4339i$, $\quad -0.2225 - 0.9749i$, $\quad 0.6235 - 0.7818i$, $\quad -0.9010 - 0.4339i$.

3. Use Durand-Kerner to find all the roots of the polynomial

$$p(x) = x^7 - x - 1.$$

Solution: We get $\zeta = 0.6171 + 0.9009i$, $-0.3636 + 0.9526i$, $-0.8099 + 0.2629i$, $-0.8099 - 0.2629i$, $-0.3636 - 0.9526i$, $0.6171 - 0.9009i$, $1.1128 + 0.0000i$.

4. Use Durand-Kerner to find all the roots of the polynomial

$$p(x) = x^8 - x - 1.$$

5. Consider now the polynomial

$$p(x) = x^6 - ax - 1,$$

where a is a real parameter. We want to investigate how the roots depend on a. For various values of $a \in [-2, 2]$, compute the roots of p and observe how they change as a changes. Can you plot the real roots as a function of a? Try to extend the range of values of a. Does anything interesting happen?

Solution: For both $a = -2$ and $a = 2$, there are two real roots, as follows:

$$a = -2: \quad \zeta_R = -1.2298, \quad 0.4928;$$

$$a = 2: \quad \zeta_R = -0.4928, \quad 1.2298.$$

Solution: What you need to do is wrap a loop structure around your Durand-Kerner routine, and add some additional code to correctly save the two real roots to make the curves shown in Fig. 3.4.

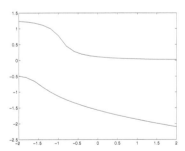

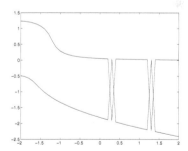

Figure 3.4   Solution curves for Problem 5, using 20 points.

Figure 3.5   Solution curves for Problem 5, using 40 points.

However, interesting things can indeed happen. If we take 40 points to construct our curves we get the picture in Fig. 3.5. What has happened?

The short answer is that polynomial roots can be very sensitive to changes in the coefficients. Specifically, in the 40 point case, some of the roots on the upper curve were converged to from the initial value used for the lower curve, and vice versa. A more sophisticated routine for the selection of the roots might be able to circumvent this problem.

6. Repeat the above for the polynomial

$$p(x) = x^6 - x - b,$$

where now $b \in [-2, 2]$ is a parameter.

7. Use MATLAB's rand function to generate a random polynomial of degree 10. (Remember to make it monic!) Use Durand-Kerner to find the roots of this polynomial, and check your results by using MATLAB's roots function.

8. Construct a random monic polynomial of degree 20, and find its roots using your Durand-Kerner routine. Compare the accuracy and average timing of your routine to the MATLAB roots function.

◁ ● ● ● ▷

3.11 SPECIAL TOPICS IN ROOT-FINDING METHODS

Exercises:

1. Consider the fixed-point iteration $x_{n+1} = 1 + e^{-x_n}$, with $x_0 = 0$. Do four steps (by hand) and apply both of the Aitken acceleration algorithms to speed up the convergence of the iteration.

   Solution: The original iteration produces $x_1 = 2.0000$, $x_2 = 1.1353$, $x_3 = 1.3213$, and $x_4 = 1.2668$. The first acceleration method produces $y_2 = 1.3963$, $y_3 = 1.2884$, and $y_4 = 1.2791$. The second acceleration method produces $z_1 = 1.3963$, $z_2 = 1.2789$, $z_2 = 1.27854$, $z_4 = 1.2785$.

2. Repeat the above for the iteration $x_{n+1} = \frac{1}{2}\ln(1 + x_n)$, using $x_0 = 1/2$.

   Solution: The basic iteration produces $x_1 = 0.2027$, $x_2 = 0.09230$, $x_3 = 0.04414$, and $x_4 = 0.02160$; the first Aitken method produces $y_2 = 0.02702$, $y_3 = 0.006906$, and $y_4 = 0.001754$; the second Aitken method produces $z_1 = 0.02702$, $z_2 = 0.1720 \times 10^{-3}$, $z_3 = 0.7392 \times 10^{-8}$, and $z_4 = 0.0000$; the exact value of the fixed point is $\alpha = 0$.

3. Consider the iteration $x_{n+1} = e^{-x}$. Show (by computational experiment) that it converges for $x_0 = 1/2$; then use both Aitken acceleration algorithms to speed up the convergence. What is the gain in terms of function calls to the iteration function?

   Solution: The second Aitken method converges to 14 digit accuracy in 5 iterations, which involves a total of 10 function calls. Neither the original iteration nor the first Aitken method converge to comparable accuracy in 20 iterations (involving 20 function calls in each case), although both are clearly close to convergence.

4. Write the second Aitken iteration in the form

$$x_{k+1} = G(x_k)$$

   Hint: Take

$$x_1 = g(x_0), \quad x_2 = g(x_1) = g(g(x_0))$$

and use this to write the updated value of x0 in Algorithm 3.5 entirely in terms of x_0, $g(x_0)$, and $g(g(x_0))$.

Solution: We have that

$$x_{n+1} = g(g(x_n)) + \gamma \frac{g(g(x_n)) - g(x_n)}{1 - \gamma},$$

where

$$\gamma = \frac{g(g(x_n)) - g(x_n)}{g(x_n) - x_n},$$

therefore,

$$x_{n+1} = g(g(x_n)) + \frac{\left(\frac{g(g(x_n)) - g(x_n)}{g(x_n) - x_n}\right)}{\left(1 - \frac{g(g(x_n)) - g(x_n)}{g(x_n) - x_n}\right)}(g(g(x_n)) - g(x_n)).$$

5. Consider the iteration

$$x_{k+1} = G(x_k),$$

where

$$G(x) = g(g(x)) + \frac{H(x)}{1 - H(x)}(g(g(x)) - g(x))$$

for

$$H(x) = \frac{g(g(x)) - g(x)}{g(x) - x}.$$

Assume that $\alpha = g(\alpha)$ is a fixed point for g.

(a) Use L'Hôpital's Rule to show that

$$H(\alpha) = g'(\alpha).$$

Solution: We have

$$
\begin{aligned}
H(\alpha) &= \lim_{x \to \alpha} \frac{g(g(x)) - g(x)}{g(x) - x} \\
&= \lim_{x \to \alpha} \frac{g'(g(x))g'(x) - g'(x)}{g'(x) - 1} \\
&= g'(\alpha) \lim_{x \to \alpha} \frac{g'(g(x)) - 1}{g'(x) - 1} \\
&= g'(\alpha) \frac{g'(\alpha) - 1}{g'(\alpha) - 1} \\
&= g'(\alpha).
\end{aligned}
$$

(b) Use Theorem 3.7 to show that the fixed point iteration for G is quadratic.

Solution: This is a direct consequence of (a) and Exercise 4. Exercise 4 implies that we can write the second Aitken algorithm as $x_{n+1} = G(x_n)$, where

$$G(x) = g(g(x)) + \frac{\left(\frac{g(g(x)) - g(x)}{g(x) - x}\right)}{\left(1 - \frac{g(g(x)) - g(x)}{g(x) - x}\right)}(g(g(x)) - g(x)).$$

Writing this in the suggested form

$$G(x) = g(g(x)) + \frac{H(x)}{1 - H(x)}(g(g(x)) - g(x)),$$

where

$$H(x) = \frac{g(g(x)) - g(x)}{g(x) - x},$$

we get that $G'(\alpha) = 0$ if $H(\alpha) = g'(\alpha)$, which is what (a) gives us.

6. Apply the chord method to each of the functions in Problem 3 of §3.1, using the midpoint of the interval as x_0. Compare your results to what you got for bisection, Newton, and/or the secant method.

Solution: Table 3.9 shows results for three of the functions. Note that two of them appear to be converging nicely, but one is cycling between two values.

Table 3.9   Solutions to Problem 3.10.6.

n	$f(x) = x^3 - 2x + 5$	$f(x) = e^x - 2$	$f(x) = x^2 - \sin x$
1	2.47368421052632	0.71306131942527	0.99684921393812
2	1.38118185096799	0.68866210956491	0.93255022118475
3	2.46066264395747	0.69409059303710	0.90477130225958
4	1.41273756576147	0.69294563585428	0.89126892528421
5	2.46661014403290	0.69319009730508	0.88438120641796
6	1.39839029026513	0.69313803554445	0.88078618703716
7	2.46412393132776	0.69314912895829	0.87888794624203
8	1.40440122265335	0.69314676542932	0.87787960220180
9	2.46521002552796	0.69314726900812	0.87734227505178
10	1.40177773347679	0.69314716171506	0.87705546338566
11	2.46474386296717	0.69314718457506	0.87690223386659
12	1.40290421128725	0.69314717970448	0.87682033179961
13	2.46494551887643	0.69314718074221	0.87677654353423
14	1.40241699456030	0.69314718052111	0.87675312931651
15	2.46485857585776	0.69314718056822	0.87674060848644
16	1.40262707133957	0.69314718055818	0.87673391267114
17	2.46489611528813	0.69314718056032	0.87673033184832
18	1.40253636927138	0.69314718055987	0.87672841685606
19	2.46487991697665	0.69314718055996	0.87672739272910
20	1.40257550786620	0.69314718055994	0.87672684503010

7. Repeat the above, except this time apply both of the Aitken acceleration algorithms to improve the convergence of the iteration. Compute the values of the ratio

$$R_n = \frac{\alpha - x_{n+1}}{(\alpha - x_n)^2},$$

where α is the best value for the root, as found by previous methods. Does the sequence of R_n values appear to be converging to a limit? What does this tell you?

Solution: Table 3.10 gives the results for the same selection of examples as in the previous problem, using Algorithm 3.5. Note that all three iterations converge to 14 digits of accuracy in only 6 iterations. The ratios R_n should converge to a limit since the accelerated methods are quadratic.

Table 3.10   Solutions to Problem 3.10.7.

n	$f(x) = x^3 - 2x + 5$	$f(x) = e^x - 2$	$f(x) = x^2 - \sin x$
1	1.95884575747166	0.69116914281208	0.91662728491450
2	2.08251651168641	0.69314676598240	0.87767216265286
3	2.09444289173521	0.69314718055993	0.87672680857003
4	2.09455147258292	0.69314718055995	0.87672621539530
5	2.09455148154233	0.69314718055995	0.87672621539506
6	2.09455148154233	0.69314718055995	0.87672621539506

8. Repeat Problem 6, but this time update the value used to compute the derivative in the chord method every other iteration. Comment on your results.

Solution: Table 3.11 gives the results for the same selection of functions as in the previous two exercises. Note that the convergence is much faster than for the pure chord method.

Table 3.11   Solutions to Problem 3.10.8.

n	$f(x) = x^3 - 2x + 5$	$f(x) = e^x - 2$	$f(x) = x^2 - \sin x$
1	2.47368421052632	0.71306131942527	0.99684921393812
2	2.15643299612282	0.69334415731550	0.89073569652677
3	2.09660460386192	0.69314719995859	0.87696243819377
4	2.09455385074497	0.69314718055995	0.87672628471248
5	2.09455148154549	0.69314718055995	0.87672621539507
6	2.09455148154233	0.69314718055995	0.87672621539506
7	2.09455148154233	0.69314718055995	0.87672621539506

9. Show that both Halley method iterations are of third order. Hint: Write them as fixed point iterations

$$x_{n+1} = g(x_n)$$

and write the iteration function as $g(x) = f(x)G(x)$. The fact that $f(\alpha) = 0$ will make the derivative computations easier.

Solution: This is a straightforward application of Theorem 3.7. For the method (3.44), we have

$$g(x) = x - \frac{2f(x)f'(x)}{2[f'(x)]^2 - f(x)f''(x)} = x - f(x)G(x)$$

for

$$G(x) = \frac{2f'(x)}{2[f'(x)]^2 - f(x)f''(x)}.$$

Then
$$g'(x) = 1 - f'(x)G(x) - f(x)G'(x).$$

Hence,
$$g'(\alpha) = 1 - f'(\alpha)G(\alpha) = 1 - f'(\alpha)\frac{2f'(\alpha)}{2[f'(\alpha)]^2} = 0.$$

Further,
$$g''(x) = -f''(x)G(x) - 2f'(x)G'(x) - f(x)G''(x),$$

so that
$$g''(\alpha) = -f''(\alpha)G(\alpha) - 2f'(\alpha)G'(\alpha),$$

which we can again show is 0.

10. Apply the Halley iteration (3.44) to $f(x) = x^2 - a$, $a > 0$, to derive a cubic method for finding the square root. Test this by using it to find $\sqrt{5}$.

Solution: The iteration is

$$x_{n+1} = \frac{x(x^2 + 3a)}{3x^2 + a}.$$

11. Repeat the above for $f(x) = x^N - a$. Test this out by finding the cubic, quartic, and quintic roots of 2, 3, 4, and 5.

Solution: The iteration for the cube root is

$$x_{n+1} = \frac{x(x^3 + 2a)}{2x^3 + a}.$$

For the quintic root it is

$$x_{n+1} = \frac{x(2x^5 + 3a)}{3x^5 + 2a}.$$

12. Apply Halley's second method to $f(x) = a - e^x$ to show that

$$x_{n+1} = x_n + 2\left(\frac{ae^{-x_n} - 1}{ae^{-x_n} + 1}\right)$$

is a cubic convergent iteration for $\alpha = \log a$. Analyze the convergence of this iteration, under the assumption that $a \in [\frac{1}{2}, 1]$. Comment on this as a practical means of computing the logarithm of an arbitrary a.

Solution: If $a \in [1/2, 1]$, then $\ln a \in [-0.7, 0]$. One can show that, for any $a \in [1/2, 1]$, the iteration function

$$g(x) = x + 2\left(\frac{ae^{-x} - 1}{ae^{-x} + 1}\right)$$

satisfies $g(x) \in [-0.7, 0]$ as well. One can then apply Theorem 3.5 to establish convergence. The only problem with this as a practical method for computing the logarithm is that it depends on having an accurate method for computing the exponential.

13. Use the secant method with $x_0 = 0$ and $x_1 = 1$ to find the root of $f(x) = 2 - e^x$, which has exact solution $\alpha = \ln 2$. Compute the ratio

$$R = \frac{\alpha - x_{n+1}}{(\alpha - x_n)^p}$$

for $p = (1 + \sqrt{5})/2$. Do we get convergence to the appropriate value?

Solution: The ratio begins to converge to the correct value but when the iteration itself converges there is massive subtractive cancellation in the computation of R, so the theoretical limit is not actually achieved. Doing the computation in a high precision environment will solve this problem.

14. Repeat the experiment in the previous exercise, but use values of p just a bit above and below the correct value. What happens?

Solution: If you don't use the correct value of p, then the ratio will either go off to infinity or zero. This isn't a result of rounding error, but is caused by having the wrong exponent in the computation of R.

15. Consider applying the secant method to find the root of the function $f(x) = 4x - \cos x$. Assume we want accuracy to within 10^{-8}. Use the secant error estimate (3.50) to show that the iteration will converge for all $x_0 \in [-2, 2]$. How many iterations will be needed for the iteration to converge?

Solution: We have

$$e_{n+1} \le \frac{1}{6} e_n e_{n-1},$$

with both $e_0 \le 4$ and $e_1 \le 4$. Therefore,

$$e_2 \le \frac{1}{6} \times 16 = \frac{8}{3};$$

$$e_3 \le \frac{1}{6} \times \frac{8}{3} \times 4 = \frac{16}{9};$$

$$e_4 \le \frac{1}{6} \times \frac{16}{9} \times \frac{8}{3} = \frac{64}{81}.$$

We get convergence to the specified accuracy in 8 iterations.

16. Verify by actual computation that your results in the previous problem were correct, i.e., apply the secant method to $f(x) = 4x - \cos x$. Do your results converge to the specified accuracy in the correct number of iterations, and does this convergence appear to be occurring for all choices of x_0?

Solution: Testing the computation for a range of values for x_0 and x_1, all in the interval $[-2, 2]$, will confirm the prediction.

17. Prove Lemma 3.1. Hint: Expand f in a Taylor series about α.

Solution: The proof follows directly from the hint.

18. Let $f(x) = 1 - xe^{1-x}$. Write this function in the form (3.57). What is $F(x)$? Use Taylor's Theorem or L'Hôpital's Rule to determine the value of $F(\alpha)$.

Solution: We have

$$f(x) = (x - 1)^2 \left(\frac{1 - xe^{1-x}}{(x - 1)^2} \right),$$

so

$$F(x) = \frac{1 - xe^{1-x}}{(x-1)^2}.$$

Applying L'Hôpital's Rule to F shows that $F(\alpha) = 1/2$.

19. For $u(x)$ as defined in (3.60), where f has a k-fold root at $x = \alpha$, show that $u(\alpha) = 0$ but $u'(\alpha) \neq 0$.

Solution: Since f is assumed to have a multiple root, we can write $f(x) = (x - \alpha)^k F(x)$ for some F. Then

$$u(x) = \frac{f'(x)}{f(x)} = \frac{k(x-\alpha)^{k-1}F(x) + (x-\alpha)^k F'(x)}{(x-\alpha)^k F(x)} = \frac{kF(x) + (x-\alpha)F'(x)}{(x-\alpha)F(x)}$$

so that $u(\alpha) = 0$ but $u'(\alpha)$ won't.

20. Use the modified Newton method (3.59) to find the root $\alpha = 1$ for the function $f(x) = 1 - xe^{1-x}$. Is the quadratic convergence recovered?

Solution: The iteration converges much more quickly than without the modification, but the computed value of the root is not as accurate as we have seen in simple root examples.

21. Let f have a double root at $x = \alpha$. Show that the Newton iteration function

$$g(x) = x - f(x)/f'(x)$$

is such that $g'(\alpha) \neq 0$. Provide all details of the calculation.

Solution: Since f has a double root we can write

$$f(x) = (x - \alpha)^2 F(x).$$

Then

$$
\begin{aligned}
g(x) &= x - \frac{(x-\alpha)^2 F(x)}{2(x-\alpha)F(x) + (x-\alpha)^2 F'(x)} = x - \frac{2(x-\alpha)F(x)}{F(x) + (x-\alpha)F'(x)} \\
&= x - (x-\alpha)\left(\frac{2F(x)}{F(x) + (x-\alpha)F'(x)}\right)
\end{aligned}
$$

so that

$$g(\alpha) = \alpha.$$

Also

$$g'(x) = 1 - (1)\left(\frac{2F(x)}{F(x) + (x-\alpha)F'(x)}\right) - (x-\alpha)\left(\frac{2F(x)}{F(x) + (x-\alpha)F'(x)}\right)'.$$

Therefore

$$g'(x) = 1 - \frac{2F(\alpha)}{F(\alpha)} - 0 = -1 \neq 0.$$

22. Using a hand calculator, carry out six iterations of the hybrid method for the function

$$f(x) = 2xe^{-15} - 2e^{-15x} + 1.$$

You should be able to match the values generated for x_2 and x_3 in the text. In addition, $x_6 = 0.04146407478711$.

Solution: $x_1 = 0.500000$, $x_2 = 0.250000$, $x_3 = 0.125069$, $x_4 = 0.0625346$, $x_5 = 0.0369239$, $x_6 = 0.0414641$

23. Write a computer code to implement the hybrid method described in §3.11.5. Apply it to each of the functions given below, using the given interval as the starting interval. On your output, be sure to indicate which iteration method was used at each step.

   (a) $f(x) = x^{1/19} - 19^{1/19}$, $[1, 100]$;

   (b) $f(x) = 2xe^{-5} - 2e^{-5x} + 1$, $[0, 1]$;

   (c) $f(x) = x^2 - (1 - x)^{20}$, $[0, 1]$;

   (d) $f(x) = 2xe^{-20} - 2e^{-20x} + 1$, $[0, 1]$;

Solution: Table 3.12 gives the results from my program, applied to the case of (d). Note that we do cycle back and forth between secant and bisection at the beginning of the iteration.

Table 3.12   Hybrid iteration for $f(x) = 2xe^{-20} - 2e^{-20x} + 1$.

n	x_n	$f(x_n)$	$x_n - x_{n-1}$	$\ln |x_n - x_{n-1}|$	a_n	b_n	Method
1	0.39540563	0.99926451	-0.3954056	-0.403	0.010	0.395	S
2	0.20270282	0.96529631	0.1927028	-0.715	0.010	0.203	B
3	0.08664329	0.64644583	0.1160595	-0.935	0.010	0.087	S
4	0.04832164	0.23912462	0.3832164×10^{-1}	-1.417	0.010	0.048	B
5	0.03786785	0.06219184	0.1045379×10^{-1}	-1.981	0.010	0.038	S
6	0.03419335	-0.00932344	0.3674505×10^{-2}	-2.435	0.034	0.038	S
7	0.03467239	0.00030061	$-0.4790448 \times 10^{-3}$	-3.320	0.034	0.035	S
8	0.03465743	0.00000139	0.1496305×10^{-4}	-4.825	0.034	0.035	S
9	0.03465736	0.00000000	0.6965962×10^{-7}	-7.157	0.035	0.035	S
10	0.03465736	0.00000000	$-0.1047069 \times 10^{-10}$	-10.980	0.035	0.035	S

I have included the Fortran code that I used to get the table (yes, Fortran is archaic):

```
c
c          Code for global root-finding routine
c          based on bisection and secant
c
           implicit real*8 (a-h,o-z)
           character*1 step,bisect,secant
           data bisect,secant / 'B','S' /
c
           f(x) = 2.0d0*x*dexp(-20.0d0)
      *         - 2.0d0*dexp(-20.0d0*x) + 1.0d0
c
           a = 0.01d0
           b = 1.00d0
           x0 = a
           x1 = b
           alpha = 0.0d0
```

```
c
      fa = f(a)
      fb = f(b)
      f0 = fa
      f1 = fb
c
      do 100 n=1,100
          xold = x
c
c     take secant step
c
          dd = f1*(x1 - x0)/(f1 - f0)
          x = x1 - f1*(x1 - x0)/(f1 - f0)
c
c     if secant prediction is outside bracketing interval,
c     do one step of bisection
c
          if((x .lt. a) .or. (x .gt. b)) then
              x = 0.5d0*(a + b)
              fx = f(x)
              if(fa*fx .lt. 0.0d0) then
                  b = x
                  fb = fx
              else
                  a = x
                  fa = fx
              endif
              x0 = a
              x1 = b
              f0 = fa
              f1 = fb
              error = xold - x
              step = bisect
          else
c
c     otherwise, update bracketing interval
c
              fx = f(x)
              if(fa*fx .lt. 0.0d0) then
                  b = x
                  fb = fx
              else
                  a = x
                  fa = fx
              endif
              error = xold - x
              step = secant
              x0 = x1
              x1 = x
              f0 = f1
              f1 = fx
          endif
      e10 = dlog10(dabs(error))
      write(6,91) n,x,fx,error,e10,a,b,step
91    format(1h ,i5,' & ',2(f12.8,' & '),e16.7,' & ',
     *       3(f8.3,' & '),1x,a1 '\\\\ \\hline')
      if(dabs(error) .le. 1.e-8) stop
100   continue
c
      stop
      end
```

24. In Problem 9 of §3.1, we introduced the regula-falsi method. In this problem we demonstrate how the hybrid method (Algorithm 3.6) is different from regula-falsi. Let

$$f(x) = e^{-4x} - \frac{1}{10}.$$

(a) Show that f has a root in the interval $[0, 5]$.

Solution: $f(0) = 1 - (1/10) > 0$, and $f(5) = e^{-20} - (1/10) < 0$.

(b) Using this interval as the starting interval, compute five iterations of regula-falsi; be sure to tabulate the new interval in addition to the new approximate root value predicted by the method at each step. A graph showing the location of each approximate root might be useful.

Solution: Table 3.13 shows five iterations of regula falsi for this function. Note that the value of a_n remains the same for the entire computation.

Table 3.13 Regula-falsi iteration for $f(x) = e^{-4x} - (1/10)$.

n	c_n	$f(c_n)$	a_n	b_n
1	4.5000	−0.1000	0	4.5000
2	4.0500	−0.1000	0	4.0500
3	3.6450	−0.1000	0	3.6450
4	3.2805	−0.1000	0	3.2805
5	2.9525	−0.1000	0	2.9525

(c) Using the same interval as the starting interval, compute five iterations of the hybrid method given in this section. Again, note the new interval as well as the new approximate root value at each step. A graphical illustration might be instructive.

Solution: Table 3.14 shows the results.

Table 3.14 Hybrid iteration for $f(x) = e^{-4x} - (1/10)$.

n	c_n	$f(c_n)$	a_n	b_n	Method
1	4.5000	−0.1000	0	4.5000	S
2	2.2500	−0.0999	0	2.2500	B
3	2.0250	−0.0997	0	2.0250	S
4	1.0125	−0.0826	0	1.0125	B
5	0.9115	−0.0739	0	0.9115	S
6	0.6222	−0.0170	0	0.6222	S

(d) Based on your results, comment on the difference between regula falsi and the hybrid method.

Solution: The obvious difference revealed by this exercise is that the addition of the bisection step speeds up the convergence compared to regula falsi, by cutting the interval in half. A more subtle difference, which is not revealed well by this example, is that in the hybrid method we use the most recent computations to compute the secant approximation, not necessarily the bracket values. This will serve to speed up the computation.

25. Write a computer program that evaluates the polynomial $p(x) = (x - 1)^5$ using the following three forms for the computation:

$$\begin{aligned}
p(x) &= (x - 1)^5, \\
p(x) &= x^5 - 5x^4 + 10x^3 - 10x^2 + 5x - 1, \\
p(x) &= -1 + x(5 + x(-10 + x(10 + x(-5 + x)))).
\end{aligned}$$

Use this to evaluate the polynomial at 400 equally spaced points on the interval $[0.998, 1.002]$. Comment on the results.

Solution: Using the first form will almost surely not produce the "cloud effect" seen in the text, while the other two probably will (exactly what happens depends a lot on the platform and language used, and the precision, of course).

◁ • • • ▷

3.12   VERY HIGH-ORDER METHODS AND THE EFFICIENCY INDEX

Exercises:

1. Write a program to employ both Chun-Neta and Super Halley to find the root of a given function. Test it on the following examples, using the given values of x_0.

   (a) $f(x) = \sin x - \frac{1}{2}x$, $x_* = 1.895494267$; $x_0 = 2, 2.5, 3$;

   (b) $f(x) = e^{x^2 + 7x - 30} - 1$, $x_* = 3$, $x_0 = 4, 5, 6$;

   (c) $f(x) = e^x \sin x + \ln(1 + x^2)$, $x_* = 1$, $x_0 = 0, -1, -2$.

Solution: What follows is a MATLAB "code segment" to do 10 iterations of the two methods; `froot` is a MATLAB function which returns the function value and its first two derivatives at the argument:

```
for k=1:10
[fx,fpx,fpxx] = froot(xsuper);
u = fx/fpx;
uL = u*fpxx/fpx;
y = xsuper - 2*u/3;
xsuper = xsuper - (1 + 0.5*uL/(1 - uL))*u;
%
[fx,fpx,fpxx] = froot(xneta);
w = xneta - fx/fpx;
[fw,dfw,d2w] = froot(w);
z = w - fw/(fpx*(1 - fw/fx)^2);
[fz,dfz,d2z] = froot(z);
%
xneta = z - fz/(fpx*(1 - (fw/fx) - (fz/fx))^2);
%
end
```

2. In §3.7 we employed (and analyzed) Newton's method as an approximator for $\sqrt{a}$ by using it to find the positive root of the function $f(x) = x^2 - a$. In this and some of the following exercises, we will try to apply our high order methods to this task.

(a) Apply Halley's method to finding the root of $f(x) = x^2 - a$. Construct an iteration function, as simplified as possible, so the iteration is $x_{n+1} = G_H(x_n)$. Verify that your construction works by testing it on $a = 0.5$, for which the exact value is $\sqrt{0.5} = 0.70710678118655$. Note the number of arithmetic operations required by your function during each iteration.

Solution: You should get

$$G_H(x) = \frac{x(x^2 + 3a)}{3x^2 + a}.$$

(b) Repeat the above for Super Halley, obtaining the iteration $x_{n+1} = G_{SH}(x_n)$ for as simple a G_{SH} as possible.

Solution: You should get

$$G_{SH}(x) = \frac{6ax^2 + 3a^2 - x^4}{8ax}.$$

3. Recall that in §3.7, we were able to restrict our attention to values in the interval $\left[\frac{1}{4}, 1\right]$, and therefore construct an initial guess by linear interpolation. Modify your codes from the above problem to reflect this, and use them to approximate $\sqrt{a}$ for $a = 0.3, 0.6$, and 0.9. In addition, write a code that uses Newton's method (or simply use your code from §3.7), and test it on the same values of a. Which method achieves full accuracy (as measured by MATLAB's sqrt function) in the fewest operations for each a?

4. Modify your codes to step from $a = 0.25$ to $a = 1$ in increments of 0.001, finding $\sqrt{a}$ for each a using each of the three methods. Thus, you will first find $\sqrt{0.25}$, then $\sqrt{0.251}$, etc., all the way to $a = 1.00$. Try to determine which root-finder gets the results fastest.

5. Repeat the above using some of the alternate initial value generators from Problems 6–8 of §3.7.

6. This is more of an essay-type question: As part of your job, you are going to be given a series of difficult root-finding problems that require highly accurate solutions. Of all the methods we have studied in this chapter, including the new methods in this section, which one would you choose? Justify your answer.

7. Think about combining one of these high-order methods with a global method as we did in §3.11.5. Can you design an algorithm that is better (faster, more efficient) than the one in that section?

◁ • • • ▷

CHAPTER 4

INTERPOLATION AND APPROXIMATION

4.1 LAGRANGE INTERPOLATION

Exercises:

1. Find the polynomial of degree 2 that interpolates at the datapoints $x_0 = 0$, $y_0 = 1$, $x_1 = 1$, $y_1 = 2$, and $x_2 = 4$, $y_2 = 2$. You should get $p_2(t) = -\frac{1}{4}t^2 + \frac{5}{4}t + 1$.

Solution:

$$
\begin{aligned}
p_2(t) &= L_0(t)y_0 + L_1(t)y_1 + L_2(t)y_2 \\
&= \frac{(t - x_1)(t - x_2)}{(0 - 1)(0 - 4)}(1) + \frac{(t - x_0)(t - x_2)}{(1 - 0)(1 - 4)}(2) + \frac{(t - x_0)(t - x_1)}{(4 - 0)(4 - 1)}(2) \\
&= \frac{(t - 1)(t - 4)}{4} - \frac{2(t - 0)(t - 4)}{3} + \frac{(t - 0)(t - 1)}{6} \\
&= -\frac{1}{4}t^2 + \frac{5}{4}t + 1.
\end{aligned}
$$

2. Find the polynomial of degree 2 that interpolates to $y = x^3$ at the nodes $x_0 = 0$, $x_1 = 1$, and $x_2 = 2$. Plot $y = x^3$ and the interpolating polynomial over the interval $[0, 2]$.

Solutions Manual to Accompany An Introduction to Numerical Methods and Analysis, Third Edition.
James F. Epperson.
© 2021 John Wiley & Sons, Inc. Published 2021 by John Wiley & Sons, Inc.

Solution:

$$
\begin{aligned}
p_2(t) &= L_0(t)y_0 + L_1(t)y_1 + L_2(t)y_2 \\
&= \frac{(t-x_1)(t-x_2)}{(0-1)(0-2)}(1) + \frac{(t-x_0)(t-x_2)}{(1-0)(1-2)}(4) + \frac{(t-x_0)(t-x_1)}{(2-0)(2-1)}(8) \\
&= 3t^2 - 2t.
\end{aligned}
$$

3. Construct the quadratic polynomial that interpolates to $y = 1/x$ at the nodes $x_0 = 1/2$, $x_1 = 3/4$, and $x_2 = 1$.

   Solution:

$$
\begin{aligned}
p_2(t) &= L_0(t)y_0 + L_1(t)y_1 + L_2(t)y_2 \\
&= \frac{(t-3/4)(t-1)}{(1/2-3/4)(1/2-1)}(2) + \frac{(t-1/2)(t-1)}{(3/4-1)(3/4-1/2)}(4/3) \\
&\quad + \frac{(t-1/2)(t-3/4)}{(1-1/2)(1-3/4)}(1) \\
&= (8/3)t^2 - 6t + (13/3).
\end{aligned}
$$

4. Construct the quadratic polynomial that interpolates to $y = \sqrt{x}$ at the nodes $x_0 = 1/4$, $x_1 = 9/16$, and $x_2 = 1$.

   Solution:

$$
\begin{aligned}
p_2(t) &= L_0(t)y_0 + L_1(t)y_1 + L_2(t)y_2 \\
&= \frac{(t-x_1)(t-x_2)}{(1/4-9/16)(1/4-1)}(1/2) + \frac{(t-x_0)(t-x_2)}{(9/16-1/4)(9/16-1)}(3/4) \\
&\quad + \frac{(t-x_0)(t-x_1)}{(1-1/4)(1-9/16)}(1) \\
&= (-32/105)t^2 + (22/21)t + (9/35).
\end{aligned}
$$

Fig. 4.1 shows the interpolate (solid line) and function (dash-dot line). Fig. 4.2 shows the error. Both plots use the circles ('o') to denote the given data.

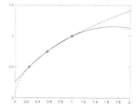
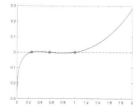

Figure 4.1   Interpolate and function for Problem 4.

Figure 4.2   Error for Problem 4.

5. For each function listed below, construct the Lagrange interpolating polynomial for the set of nodes specified. Plot both the function and the interpolating polynomial, and also the error, on the interval defined by the nodes.

(a) $f(x) = \ln x$, $x_i = 1, \frac{3}{2}, 2$;

(b) $f(x) = \sqrt{x}$, $x_i = 0, 1, 4$;

(c) $f(x) = \log_2 x$, $x_i = 1, 2, 4$;

(d) $f(x) = \sin \pi x$, $x_i = -1, 0, 1$.

Solution:

(a) $p_2(x) = -0.235566072x^2 + 1.39984540x - 1.164279325$;

(b) $p_2(x) = (7x - x^2)/6$;

(c) $p_2(x) = -(1/6)x^2 + (3/2)x - (4/3)$;

(d) $p_2(x) = 0$.

6. Find the polynomial of degree 3 that interpolates $y = x^3$ at the nodes $x_0 = 0$, $x_1 = 1$, $x_2 = 2$, and $x_3 = 3$. (Simplify your interpolating polynomial as much as possible.) Hint: This is easy if you think about the implications of the uniqueness of the interpolating polynomial.

Solution: You should get $p_3(x) = x^3$; interpolation to a polynomial reproduces the polynomial if enough nodes are used.

7. Construct the Lagrange interpolating polynomial to the function $f(x) = x^2 + 2x$, using the nodes $x_0 = 0$, $x_1 = 1$, $x_2 = -2$. Repeat, using the nodes $x_0 = 2$, $x_1 = 1$, $x_2 = -1$. (For both sets of nodes, simplify your interpolating polynomial as much as possible.) Comment on your results, especially in light of the uniqueness part of Theorem 4.1, and then write down the interpolating polynomial for interpolating to $f(x) = x^3 + 2x^2 + 3x + 1$ at the nodes $x_0 = 0$, $x_1 = 1$, $x_2 = 2$, $x_3 = 3$, $x_4 = 4$, and $x_5 = 5$. Hint: You should be able to do this last part without doing any computations.

Solution: In all three cases the interpolating polynomial will be identical to the original polynomial.

8. Let f be a polynomial of degree $\leq n$, and let p_n be a polynomial interpolant to f, at the $n + 1$ distinct nodes $x_0, x_1, \ldots, x_n$. Prove that $p_n(x) = f(x)$ for all x, i.e., that interpolating to a polynomial will reproduce the polynomial, if you use enough nodes. Hint: Consider the uniqueness part of Theorem 4.1.

Solution: Let $q(x) = f(x) - p_n(x)$. Since both f and p_n are polynomials of degree $\leq n$, so is q. But $q(x_k) = 0$ for the $n + 1$ nodes x_k. The only way a polynomial of degree $\leq n$ can have $n + 1$ roots is if it is identically zero, hence $f(x) = p_n(x)$ for all x.

9. Let p be a polynomial of degree $\leq n$. Use the uniqueness of the interpolating polynomial to show that we can write

$$p(x) = \sum_{i=0}^{n} L_i^{(n)}(x)p(x_i)$$

for any distinct set of nodes $x_0, x_1, \ldots, x_n$. Hint: See the previous problem.

Solution: Define

$$q(x) = p(x) - \sum_{i=0}^{n} L_i^{(n)}(x)p(x_i).$$

Since everything on the right side is a polynomial of degree $\leq n$, so is q. But, again, $q(x_k) = 0$ for the $n + 1$ nodes x_k, therefore $q(x) = 0$ for all x, which proves the desired result.

10. Show that

$$\sum_{i=0}^{n} L_i^{(n)}(x) = 1$$

for any set of distinct nodes x_k, $0 \leq k \leq n$. Hint: This does not require *any* computation with the sum, but rather a perceptive choice of polynomial p in the previous problem.

Solution: Let $p(x) = 1$, and note that this is a polynomial of degree 0. By the previous problem, we have that

$$1 = p(x) = \sum_{i=0}^{n} L_i^{(n)}(x)p(x_i) = \sum_{i=0}^{n} L_i^{(n)}(x),$$

and we are done.

$$\triangleleft \bullet \bullet \bullet \triangleright$$

4.2   NEWTON INTERPOLATION AND DIVIDED DIFFERENCES

Exercises:

1. Construct the polynomial of degree 3 that interpolates to the data $x_0 = 1$, $y_0 = 1$, $x_1 = 2$, $y_1 = 1/2$, $x_2 = 4$, $y_2 = 1/4$, and $x_3 = 3$, $y_3 = 1/3$. You should get $p(t) = (50 - 35t + 10t^2 - t^3)/24$.

   Solution: The divided difference coefficients are 1, $-1/2$, $1/8$, and $-1/24$, so the polynomial is

   $$p(t) = 1 - (1/2)(x - 1) + (1/8)(x - 1)(x - 2) - (1/24)(x - 1)(x - 2)(x - 4),$$

   which simplifies as indicated.

2. For each function listed below, use divided difference tables to construct the Newton interpolating polynomial for the set of nodes specified. Plot both the function and the interpolating polynomial, and also the error, on the interval defined by the nodes.

   (a) $f(x) = \sqrt{x}$, $x_i = 0, 1, 4$;

       Solution: The divided difference coefficients are $a_0 = 0$, $a_1 = 1$, and $a_2 = -1/6$.

   (b) $f(x) = \ln x$, $x_i = 1, \frac{3}{2}, 2$;

   (c) $f(x) = \sin \pi x$, $x_i = 0, \frac{1}{4}, \frac{1}{2}, \frac{3}{4}, 1$;

   (d) $f(x) = \log_2 x$, $x_i = 1, 2, 4$;

Solution: The divided difference coefficients are $a_0 = 0$, $a_1 = 1$, and $a_2 = 1/6$, so the polynomial is

$$p(t) = 0 + (1)(x - 1) + (1/6)(x - 1)(x - 2),$$

which simplifies to
$$p(t) = (1/6)(x^2 + 3x - 4).$$

(e) $f(x) = \sin \pi x$, $x_i = -1, 0, 1$.

3. Let $f(x) = e^x$. Define $p_n(x)$ to be the Newton interpolating polynomial for $f(x)$, using $n + 1$ equally spaced nodes on the interval $[-1, 1]$. Thus we are taking higher and higher degree polynomial approximations to the exponential function. Write a program that computes $p_n(x)$ for $n = 2, 4, 8, 16, 32$, and which samples the error $f(x) - p_n(x)$ at 501 equally spaced points on $[-1, 1]$. Record the maximum error as found by the sampling, as a function of n, i.e., define E_n as

$$E_n = \max_{0 \le k \le 500} |f(t_k) - p_n(t_k)|,$$

where $t_k = -1 + 2k/500$, and plot E_n vs. n.

Solution: Using MATLAB on a low-end personal computer, the author got

$$
\begin{aligned}
E_2 &= 7.8525 \times 10^{-2}, \quad E_4 = 1.1244 \times 10^{-3}, \quad E_8 = 5.8000 \times 10^{-8}, \\
E_{16} &= 1.5321 \times 10^{-14}, E_{32} = 3.6471 \times 10^{-10}.
\end{aligned}
$$

Note that the error started to actually *increase* as we added more nodes. This is explained, in part, by the material in §4.11.1.

4. In §3.7 we used linear interpolation to construct an initial guess for Newton's method as a means of approximating $\sqrt{a}$. Construct the quadratic polynomial that interpolates the square root function at the nodes $x_0 = \frac{1}{4}, x_1 = \frac{4}{9}, x_2 = 1$. Plot the error between p_2 and $\sqrt{x}$ over the interval $[\frac{1}{4}, 1]$ and try to estimate the worst error. What impact will this have on the use of Newton's method for finding $\sqrt{a}$?

Solution: The polynomial is

$$p(t) = -\frac{12}{35} t^2 + \frac{23}{21} t + \frac{26}{105}.$$

The worst absolute error, $|\sqrt{t} - p(t)|$, appears to be, from looking at a plot, about 0.01. This is significantly smaller than the error due to linear interpolation, so the Newton iteration for computing $\sqrt{a}$ will converge faster. See some of the exercises in §3.7.

5. Prove Corollary 4.1.

Solution: A straight-forward inductive argument works.

6. Write and test a computer code that takes a given set of nodes, function values, and corresponding divided difference coefficients, and computes new divided difference coefficients for new nodes and function values. Test it by recursively computing interpolating polynomials that approximate $f(x) = e^x$ on the interval $[0, 1]$.

Solution: A set of MATLAB scripts that accomplish this are given below. The first script (`polyterp`) is the "main program." The second one (`divdif`) computes divided difference coefficients. The third one (`intval`) evaluates an interpolating polynomial from nodal data and divided difference coefficients.

```
%
n = input('Initial degree? ');
m = input('Maximum degree? ');
np = n + 1;
mp = m + 1;
%
xn = [0:n]/n
yn = exp(xn);
x = [0:500]/500;
a = divdif(xn,yn);
y = intval(x,xn,a);
figure(1)
plot(x,y)
figure(2)
plot(x,y - exp(x));
max(abs(y - exp(x)))
%
for k=np:mp
    xx = input('New node value? ');
    yy = exp(xx);
    pn = intval(xx,xn,a);
    w = xx - xn;
    w = prod(w);
    xn = [xn xx];
    yn = [yn yy];
    aa = (yy - pn)/w;
    a = [a aa];
    y = intval(x,xn,a);
    max(abs(y - exp(x)))
    check = sum(abs(yn - intval(xn,xn,a)))
end
%%%%%%%%%%%%%%%%%%%%%%%%%%%%%%%%%%%%%%%%%%%

function a = divdif(x,y)
    n = length(x)-1;
    a = y;
    for i=1:n
        for j=n:(-1):i
            a(j+1) = (a(j+1) - a(j))/(x(j+1) - x(j-i+1));
        end
    end
%%%%%%%%%%%%%%%%%%%%%%%%%%%%%%%%%%%%%%%%%%%
```

```
function y = intval(x,xx,d)
%
%      evaluates polynomial using divided difference
%      coefficients d, and interpolation data xx
%
       m = length(x);
       n = length(d);
%
       y = d(n)*ones(1,m);
       for i=(n-1):(-1):1
           y = d(i) + (x - xx(i)*ones(1,m)).*y;
       end
```

7. In 1973, the horse Secretariat became the first (and, so far, only) winner of the Kentucky Derby to finish the race in less than 2 minutes, running the $1\frac{1}{4}$ mile distance in $1:59.4$ seconds[1]. Remarkably, he ran each quarter mile *faster* than the previous one, as the Table 4.1 of data shows. Here, t is the elapsed time (in seconds)

Table 4.1   Data for Problem 7.

x	0.0	0.25	0.50	0.75	1.00	1.25
t	0.0	25.0	49.4	73.0	96.4	119.4

since the race began and x is the distance (in miles) that Secretariat has traveled.

(a) Find the cubic polynomial that interpolates this data at $x = 0, 1/2, 3/4, 5/4$.
   Solution:

$$T(x) = p_3(x) = 2.986666666\, x^3 - 9.599999999\, x^2 + 102.8533333\, x.$$

(b) Use this polynomial to estimate Secretariat's speed at the finish of the race, by finding $p_3'(5/4)$.
   Solution: We need to use a little calculus here. We have $t = P_3(x)$, but the speed of the horse comes from computing dx/dt, *not* dt/dx. But we know that

$$\frac{dx}{dt} = \frac{1}{\frac{dt}{dx}},$$

so the speed of the horse at $x = 5/4$ is given by

$$S = \frac{1}{p_3'(5/4)} = 0.1076967261 \times 10^{-1}.$$

Unfortunately, this is in miles per second and we wanted miles per hour. To convert, we simply multiply by 3,600 seconds per hour to get

$$S = 38.77082140.$$

---

[1] During the final preparation of the First Edition of the text, in May, 2001, the horse Monarchos won the Kentucky Derby in a time of $1:59.8$ seconds, becoming the second winner of the race to finish in less than two minutes. Ironically, the only other horse to run the Derby in less than two minutes was Sham, who finished second to Secretariat. As I work to prepare the Third Edition, no other horses have run the Kentucky Derby in less than 2:00 minutes.

(c) Find the quintic polynomial that interpolates the entire dataset.

Solution:

$$
\begin{aligned}
p_5(x) &= -13.65333333\,x^5 + 42.66666667\,x^4 - 44.80000001\,x^3 \\
&+ 13.33333333\,x^2 + 98.85333333\,x.
\end{aligned}
$$

(d) Use the quintic polynomial to estimate Secretariat's speed at the finish line.

Solution: This time we get $S = 40.51620641$ miles per hour.

8. The data in Table 4.2 (Table 4.7 in the text) gives the actual thermal conductivity data for the element mercury. Use Newton interpolation and the data for $300°$, $500°$, and $700°$ to construct a quadratic interpolate for this data. How well does it predict the values at $400°$ and $600°$?

Table 4.2   Data for Problem 8.

Temperature ($°K$), u	300	400	500	600	700
Conductivity (watts/cm $°K$), k	0.084	0.098	0.109	0.12	0.127

Solution: The polynomial is

$$
p_2(T) = -0.0000000875\,T^2 + 0.000195\,T + 0.033375.
$$

Since $p_2(400) = 0.97375 \times 10^{-1}$ and $p_2(600) = 0.118875$ it is apparent that the interpolating polynomial does a good job of matching the data.

9. The gamma function, denoted by $\Gamma(x)$, is an important special function in probability, combinatorics, and many other areas of applied mathematics. Because it can be shown that $\Gamma(n+1) = n!$, the gamma function is considered a generalization of the factorial function to non-integer arguments. Table 4.3 (Table 4.8 in the text) gives the values of $\Gamma(x)$ on the interval $[1, 2]$. Use these to construct the fifth degree polynomial based on the nodes $x = 1, 1.2, 1.4, 1.6, 1.8, 2.0$, and then use this polynomial to estimate the values at $x = 1.1, 1.3, 1.5, 1.7, 1.9$. Plot your polynomial and compare it to the intrinsic gamma function on your computing system or calculator.

Solution: We get

$$
\begin{aligned}
p_5 &= -0.09270377605x^5 + 0.8632574013x^4 - 3.199846710x^3 + 6.287123228x^2 \\
&- 6.537486675x + 3.679656532.
\end{aligned}
$$

and

$$
p_5(1.1) = 0.951439127, \quad p_5(1.3) = 0.897445725, \quad p_5(1.5) = 0.886242424,
$$

$$
p_5(1.7) = 0.908619509, \quad p_5(1.9) = 0.961817632,
$$

which is pretty good agreement.

The plots are in Figs. 4.3–4.4, and confirm the accuracy of the approximation. (Figure 4.3 actually shows both the interpolate and $\Gamma(x)$, they are visually indistinguishable.)

Table 4.3   Table of $\Gamma(x)$ values.

x	$\Gamma(x)$
1.00	1.0000000000
1.10	0.9513507699
1.20	0.9181687424
1.30	0.8974706963
1.40	0.8872638175
1.50	0.8862269255
1.60	0.8935153493
1.70	0.9086387329
1.80	0.9313837710
1.90	0.9617658319
2.00	1.0000000000

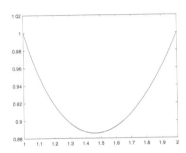

Figure 4.3   Approximation to $\Gamma(x)$.   **Figure 4.4**   Error in approximation to $\Gamma(x)$.

10. The error function, which we saw briefly in Chapters 1 and 2, is another important special function in applied mathematics, with applications to probability theory and the solution of heat conduction problems. The formal definition of the error function is as follows.

$$\text{erf}(x) = \frac{2}{\sqrt{\pi}} \int_0^x e^{-t^2}\,dt.$$

Table 4.4 (Table 4.9 in the text) gives values of $\text{erf}(x)$ in increments of 0.1, over the interval $[0, 1]$.

(a) Construct the quadratic interpolating polynomial to the error function using the data at the nodes $x_0 = 0$, $x_1 = 0.5$, and $x_2 = 1.0$. Plot the polynomial and the data in the table and comment on the observed accuracy.

   Solution: The divided difference coefficients are

   $$a_0 = 0, \quad a_1 = 1.0410, \quad a_2 = -3.9660 \times 10^{-1}$$

   and the plot of the interpolating polynomial is given in Figure 4.5. The error is given in Figure 4.6. This is decent agreement.

(b) Repeat the above, but this time construct the cubic interpolating polynomial using the nodes $x_0 = 0.0$, $x_2 = 0.3$, $x_2 = 0.7$, and $x_3 = 1.0$.

Table 4.4   Table of $\mathrm{erf}(x)$ values for Problem 10.

x	$\mathrm{erf}(x)$
0.0	0.00000000000000
0.1	0.11246291601828
0.2	0.22270258921048
0.3	0.32862675945913
0.4	0.42839235504667
0.5	0.52049987781305
0.6	0.60385609084793
0.7	0.67780119383742
0.8	0.74210096470766
0.9	0.79690821242283
1.0	0.84270079294971

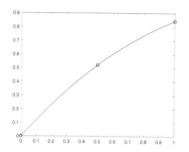
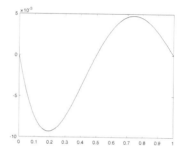

Figure 4.5   Solution plot for Exercise 4.2.10a.   **Figure 4.6**   Error plot for Exercise 4.2.10a.

Solution: The divided difference coefficients are now

$$a_0 = 0, \quad a_1 = 1.0954, \quad a_2 = -3.1784 \times 10^{-1}, \quad a_3 = -1.4398 \times 10^{-2}$$

and the plot of the interpolating polynomial is given in Figure 4.7. This is slightly better than in Part (a).

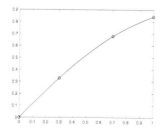
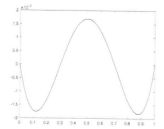

Figure 4.7   Solution plot for Exercise 4.2.10b.   **Figure 4.8**   Error plot for Exercise 4.2.10b.

11. As steam is heated up, the pressure it generates is increased. Over the temperature range $[220, 300]$ ($^\circ$ F) the pressure, in pounds per square inch, is given in Table 4.5.[2] (This is Table 4.10 in the text.)

T	220	230	240	250	260	270	280	290	300
P	17.188	20.78	24.97	29.82	35.42	41.85	49.18	57.53	66.98

Table 4.5   Temperature-pressure values for steam; Problem 11.

(a) Construct the quadratic interpolating polynomial to this data at the nodes $T_0 = 220$, $T_1 = 260$, and $T_2 = 300$. Plot the polynomial and the data in the table and comment on the observed accuracy.

Solution: The divided difference coefficients are

$$a_0 = 17.1880, \quad a_1 = 0.4558, \quad a_2 = 0.0042$$

and the plot of the interpolating polynomial is given in Figure 4.9. This is very good agreement.

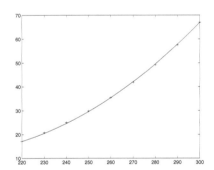

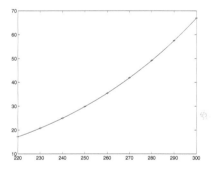

Figure 4.9   Solution plot for Exercise 4.2.11a.

Figure 4.10   Solution plot for Exercise 4.2.11b.

(b) Repeat the above, but this time construct the quartic interpolating polynomial using the nodes $T_0 = 220$, $T_1 = 240$, $T_2 = 260$, $T_3 = 280$, and $T_4 = 300$.

Solution: The divided difference coefficients are

$$\begin{aligned} a_0 &= 17.188, \quad a_1 = 3.8910 \times 10^{-1}, \quad a_2 = 3.3350 \times 10^{-3}, \\ a_3 &= 1.3375 \times 10^{-5}, \quad a_4 = 2.2917 \times 10^{-8} \end{aligned}$$

and the plot of the interpolating polynomial is given in Figure 4.10.

(c) Which of the two polynomials would you think is best to use to get values for $P(T)$ that are not on the table?

Solution: Both plots look very good, actually.

---

[2]Taken from tables in *Introduction to Thermodynamics: Classical and Statistical*, by Sonntag and Van Wylen, John Wiley & Sons, New York, 1971.

12. Similar data for gaseous ammonia are given in Table 4.6 (Table 4.11 in the text).

Table 4.6 Temperature-pressure values for gaseous ammonia; Problem 12.

T	0	5	10	15	20	25	30	35	40
P	30.42	34.27	38.51	43.14	48.21	53.73	59.74	66.26	73.32

(a) Construct the quadratic interpolating polynomial to this data at the nodes $T_0 = 0$, $T_1 = 20$, and $T_2 = 40$. Plot the polynomial and the data in the table and comment on the observed accuracy.

Solution: The divided difference coefficients are

$$a_0 = 30.420, \quad a_1 = 8.8950 \times 10^{-1}, \quad a_2 = 9.1500 \times 10^{-3}.$$

(b) Repeat the above, but this time construct the quartic interpolating polynomial using the nodes $T_0 = 0, T_1 = 10, T_2 = 20, T_3 = 30$, and $T_4 = 40$.

Solution: The divided difference coefficients are

$$a_0 = 30.420, \quad a_1 = 8.0900 \times 10^{-1}, \quad a_2 = 8.0500 \times 10^{-3}, \quad a_3 = 3.6667 \times 10^{-5},$$
$$a_4 = 1.7279 \times 10^{-20}.$$

(c) Which of the two polynomials would you think is better for getting accurate values for $P(T)$ that are not on the table?

13. In Problem 8 of §3.1 and Problem 8 of §3.8 we looked at the motion of a liquid/solid interface under a simplified model of the physics involved, in which the interface moved according to

$$x = 2\beta\sqrt{t}$$

for $\beta = \alpha/\sqrt{k}$. Here k is a material property and α is the root of $f(z) = \theta e^{-z^2} - z\,\text{erf}(z)$, where θ also depends on material properties. Figure 4.11 shows a plot of α vs. $\log_{10}\theta$, based on finding the root of $f(z)$. Some of the data used to create this curve are given in Table 4.7 (Table 4.12 in the text).

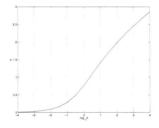

Figure 4.11 Figure for Problem 13.

(a) Use the data at the nodes $\{-6, -4, -2, 0, 2, 4, 6\}$ to construct an interpolating polynomial for this table. Plot the polynomial and compare to the actual

Table 4.7   Data for Problem 13.

$\log_{10} \theta$	α
−6.0000	0.944138×10^{-3}
−5.0000	0.298500×10^{-2}
−4.0000	0.941277×10^{-2}
−3.0000	0.297451×10^{-1}
−2.0000	0.938511×10^{-1}
−1.0000	0.289450
0.0000	0.767736
1.0000	1.41492
2.0000	1.98151
3.0000	2.45183
4.0000	2.85669
5.0000	3.21632
6.0000	3.54269

graph in Figure 4.11. Use the polynomial to compute values at $\log_{10} \theta = -5, -3, \ldots, 3, 5$. How do these compare to the actual values from the table?

Solution: The divided difference coefficients are

$$a_0 = 9.4414 \times 10^{-4}, \quad a_1 = 4.2343 \times 10^{-3}, \quad a_2 = 9.4962 \times 10^{-3}, \quad a_3 = 1.0697 \times 10^{-2},$$

$$a_4 = -1.4662 \times 10^{-3}, \quad a_5 = -6.9243 \times 10^{-5}, \quad a_6 = 4.6066 \times 10^{-5}.$$

We get $p_6(x) = 0.0001600x^6 + 0.00066298x^5 - 0.007638x^4 - 0.03204x^3 + 0.09548x^2 + 0.5895x + 0.7677$, for $x = \log_{10} \theta$.

(b) Compute the higher-degree Newton polynomial based on the entire table of data. Plot this polynomial and compare it to the one generated using only part of the data.

Solution: Figure 4.12 shows the nodes and both interpolating polynomials. Note that the interpolations have much more "wiggle" to them than is present in the raw data.

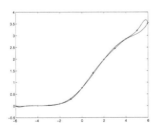

Figure 4.12   Solution plot for Exercise 4.2.13.

14. Write a computer program to construct the Newton interpolating polynomial to $f(x) = \sqrt{x}$ using equally spaced nodes on the interval $[0, 1]$. Plot the error $f(x) - p_n(x)$ for $n = 4, 8, 16$ and comment on what you get.

Solution: Figure 4.13 shows the three error plots. Note the large "spike" in the error near $x = 0$. §4.3 will explain why this occurs.

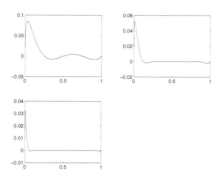

Figure 4.13   Solution plot for Exercise 4.2.14.

◁ • • • ▷

4.3 INTERPOLATION ERROR

Exercises:

1. What is the error in quadratic interpolation to $f(x) = \sqrt{x}$, using equally spaced nodes on the interval $[\frac{1}{4}, 1]$?

   Solution: We have

   $$
   \begin{aligned}
   |f(x) - p_2(x)| &\leq \frac{1}{9\sqrt{3}}(3/8)^3 \max_{t \in [1/4,1]} |(3/8)t^{-5/2}| \\
   &= \frac{1}{9\sqrt{3}} \times \frac{81}{4096} \times 4^{5/2} = 0.4059494081 \times 10^{-1}.
   \end{aligned}
   $$

2. Repeat the above for $f(x) = x^{-1}$ on $[\frac{1}{2}, 1]$.

   Solution: We have

   $$
   \begin{aligned}
   |f(x) - p_2(x)| &\leq \frac{1}{9\sqrt{3}}(1/4)^3 \max_{t \in [1/2,1]} |6x^{-4}| \\
   &= \frac{1}{9\sqrt{3}} \times \frac{1}{64} \times 6 \times 16 = 0.9622504490 \times 10^{-1}.
   \end{aligned}
   $$

3. Repeat the previous two problems, using cubic interpolation.

   Solution: This time we have

   $$
   \begin{aligned}
   |\sqrt{x} - p_3(x)| &\leq \frac{1}{24}(1/4)^4 \max_{t \in [1/4,1]} |(15/16)x^{-7/2}| \\
   &= \frac{1}{24} \times \frac{1}{256} \times \frac{15}{16} \times 4^{7/2} = 0.1953125 \times 10^{-1}
   \end{aligned}
   $$

and

$$|x^{-1}-p_3(x)| \le \frac{1}{24}(1/6)^4 \max_{t\in[1/2,1]} |24x^{-5}| = \frac{1}{24}\times\frac{1}{36^2}\times 24\times 32 = 0.2469135802\times 10^{-1}.$$

4. Show that the error in third degree polynomial interpolation satisfies

$$\|f - p_3\|_\infty \le \frac{1}{24}h^4\|f^{(4)}\|_\infty,$$

if the nodes x_0, x_1, x_2, x_3 are equally spaced, with $x_i - x_{i-1} = h$. Hint: Use the change of variable $t = x - x_1 - \frac{1}{2}h$.

Solution: We have

$$f(x) - p_4(x) = \frac{w_4(x)}{24}f^{(4)}(\xi_x),$$

for some value ξ_x in the interval defined by the nodes and x. Therefore,

$$\|f - p_3\|_\infty \le \frac{1}{24}\max_{x\in[x_0,x_3]}|w_4(x)|\|f^{(4)}\|_\infty,$$

where we have assumed that the nodes are ordered $x_0 < x_1 < x_2 < x_3$. The change of variable reduces

$$W = \max_{x\in[x_0,x_3]}|w_4(x)|$$

to

$$W = \max_{t\in[-3h/2,3h/2]}|(t^2 - 9h^2/4)(t^2 - h^2/4)|$$

and ordinary calculus shows that $W = h^4$.

5. Show that the error in polynomial interpolation using six equally spaced points (quintic interpolation) satisfies

$$\|f - p_5\|_\infty \le Ch^6\|f^{(6)}\|_\infty.$$

where $C \approx 0.0235$. Hint: See previous problem.

Solution: The basic outline from the previous exercise shows us that

$$\|f - p_5\|_\infty \le \frac{1}{720}\max_{t\in[-5h/2,5h/2]}|(t^2 - 25h^2/4)(t^2 - 9h^2/4)(t^2 - h^2/4)|\|f^{(6)}\|_\infty.$$

Applying ordinary calculus to the problem of finding

$$W = \max_{t\in[-5h/2,5h/2]}|(t^2 - 25h^2/4)(t^2 - 9h^2/4)(t^2 - h^2/4)|$$

shows that $W = 16.90089433h^6$, thus $C = 16.90089433/720 = 0.0235$.

6. Generalize the derivation of the error bound (4.12) for quadratic interpolation to the case where the nodes are *not* equally spaced. Take $x_1 - x_0 = h$ and $x_2 - x_1 = \theta h$ for some $\theta > 0$.

Solution: The issue is obtaining an upper bound for

$$|w_2(x)| = |(x - x_0)(x - x_1)|.$$

Ordinary calculus, coupled with the same kind of variable shift as used in the text, can be used to show that

$$|w_2(x)| \leq (h^3/27)g(\theta),$$

where

$$g(\theta) = -\left(\theta - 2 + \sqrt{\theta^2 - \theta + 1}\right)\left(\theta + 1 + \sqrt{\theta^2 - \theta + 1}\right)\left(2\theta - 1 - \sqrt{\theta^2 - \theta + 1}\right)$$

for $\theta \in [0, 1/2]$, and

$$g(\theta) = -\left(\theta - 2 - \sqrt{\theta^2 - \theta + 1}\right)\left(\theta + 1 - \sqrt{\theta^2 - \theta + 1}\right)\left(2\theta - 1 + \sqrt{\theta^2 - \theta + 1}\right)$$

for $\theta \in [1/2, 1]$.

This looks messy, but the graph of g shows a simple V-shaped curve, ranging from 4 at $\theta = 0$ to slightly more than 1 at $\theta = 1/2$, and then back up as θ continues to increase.

7. Apply your result for the previous problem to the error in quadratic interpolation to $f(x) = \sqrt{x}$ using the nodes $x_0 = \frac{1}{4}$, $x_1 = \frac{9}{16}$, and $x_2 = 1$.

   Solution: We have that $h = x_1 - x_0 = 5/16$, and $\theta = 7/5$. Thus, the error is bounded according to

   $$\begin{aligned} |\sqrt{x} - p_2(x)| &\leq (h^3/27)g(\theta) \max_{t \in [0,1]} |(-1/4)t^{-3/2}| \\ &= (5/16)^3 \times (1/27) \times 6.488878750 \times (1/4) \times 4^{3/2} \\ &= 0.1466850846 \times 10^{-1}. \end{aligned}$$

8. If we want to use a table of exponential values to interpolate the exponential function on the interval $[-1, 1]$, how many points are needed to guarantee 10^{-6} accuracy with linear interpolation? Quadratic interpolation?

   Solution: We have, for the linear case,

   $$\|f - p_n\|_\infty \leq \frac{1}{8}h^2\|f''\|_\infty = \frac{e}{8}h^2.$$

   Therefore we get 10^{-6} accuracy when $h \leq 0.1715527770 \times 10^{-2}$, which implies $n \geq 1166$ points on the interval $[-1, 1]$.

   For the quadratic case, we have

   $$\|f - p_n\|_\infty \leq \frac{1}{9\sqrt{3}}h^3\|f'''\|_\infty = \frac{e}{9\sqrt{3}}h^3.$$

   Therefore we get 10^{-6} accuracy when $h \leq 0.1789930706 \times 10^{-1}$, which implies $n \geq 112$ points.

9. If we want to use a table of values to interpolate the error function on the interval $[0, 5]$, how many points are needed to get 10^{-6} accuracy using linear interpolation? Quadratic interpolation? Would it make sense to use one grid spacing on, say, $[0, 1]$, and another one on $[1, 5]$? Explain.

Solution: For the linear case, we have

$$\|f - p_n\|_\infty \le \frac{1}{8}h^2 \|f''\|_\infty = \frac{h^2}{2\sqrt{pi}} \, (1).$$

Therefore we get 10^{-6} accuracy when $h \le 0.1882792528 \times 10^{-2}$, which implies $n \ge 2656$ points on the interval $[0, 5]$

For the quadratic case, we have

$$\|f - p_n\|_\infty \le \frac{h^3}{9\sqrt{3}} \|f'''\|_\infty = \frac{2.256758334 h^3}{9\sqrt{3}}.$$

Therefore we get 10^{-6} accuracy when $h \le 0.1904463677 \times 10^{-1}$, which implies $n \ge 263$ points on the interval $[0, 5]$

Breaking the table up into two regions $[0, 1]$ and $[1, 5]$ makes sense because the error function is nearly flat for large arguments, and so we can get away with a much larger h on $[1, 5]$ and probably many fewer overall points.

10. If we want to use a table of values to interpolate the sine function on the interval $[0, \pi]$, how many points are needed for 10^{-6} accuracy with linear interpolation? Quadratic interpolation? Cubic interpolation?

    Solution: We get $n \ge 1,111$ for the linear case, $n \ge 144$ for the quadratic case, $n \ge 45$ for the cubic case.

11. Let's return to the computation of the natural logarithm. Consider a computer which stores numbers in the form $z = f \cdot 2^\beta$, where $\frac{1}{2} \le f \le 1$. We want to consider using this, in conjunction with interpolation ideas, to compute the natural logarithm function.

    (a) Using piecewise linear interpolation over a grid of equally spaced points, how many table entries would be required to accurately approximate $\ln z$ to within 10^{-14}?

    (b) Repeat the above, using piecewise quadratic interpolation.

    (c) Repeat it again, using piecewise cubic interpolation.

    Explain, in a brief essay, the importance here of restricting the domain of z to the interval $[\frac{1}{2}, 1]$.

    Solution: For the linear case, we have

    $$|\ln x - p_1(x)| \le (1/8)h^2 \max_{t \in [1/2, 1]} |-1/t^2| = h^2/2,$$

    so we need $h \le 0.1414213562 \times 10^{-6}$ and hence $n \ge 3,535,533.907$ points. For the quadratic case, we have

    $$|\ln x - p_2(x)| \le (1/(9\sqrt{3}))h^3 \max_{t \in [1/2, 1]} |2/t^3| = 16h^3/(9\sqrt{3})$$

    so we need $h \le 0.2135802307 \times 10^{-4}$ and hence $n \ge 23,410.40640$ points. Finally, for the cubic case, we have

    $$|\ln x - p_3(x)| \le (1/24)h^4 \max_{t \in [1/2, 1]} |-6/t^4| = 96h^4/24 = 4h^4$$

so we need $h \leq 0.2236067977 \times 10^{-3}$ and hence $n \geq 2,236.067978$ points.

12. Assume that, for any real c,

$$\lim_{n \to \infty} \frac{c^n}{n!} = 0.$$

Use this to prove that, if p_n is the polynomial interpolate to $f(x) = \sin x$ on the interval $[a, b]$, using *any* distribution of distinct nodes x_k, $0 \leq k \leq n$, then $\|f - p_n\|_\infty \to 0$ as $n \to \infty$. Hint: Can we justify $|w_n(x)| \leq c^{n+1}$ for some c?

Solution: We have

$$\|f - p_n\|_\infty \leq \frac{\|w_n\|_\infty}{(n+1)!} \|f^{(n+1)}\|_\infty \leq \frac{\|w_n\|_\infty}{(n+1)!},$$

but

$$\|w_n\|_\infty = \max_{x \in [a,b]} \prod_{k=0}^{n} |x - x_n| \leq \prod_{k=0}^{n} |b - a| = (b - a)^{n+1}.$$

Therefore,

$$\|f - p_n\|_\infty \leq \frac{(b-a)^{n+1}}{(n+1)!} \to 0,$$

according to the assumed hypothesis.

13. Can you repeat the above for $f(x) = e^x$? Why/why not?

Solution: A similar argument will work for the exponential function, since the derivative term is again bounded by a constant.

14. Define the norms

$$\|f\|_\infty = \max_{x \in [0,1]} |f(x)|$$

and

$$\|f\|_2 = \left(\int_0^1 [f(x)]^2 dx \right)^{1/2}.$$

Compute $\|f\|_\infty$ and $\|f\|_2$ for the following list of functions:

(a) $\sin n\pi x$, n integer;

(b) e^{-ax}, $a > 0$;

(c) $\sqrt{x} e^{-ax^2}$, $a > 0$;
    Solution: If $a > 1/4$ then $\|f\|_\infty = (\sqrt{2}/2)e^{-1/4}a^{-1/4}$; if $a \leq 1/4$, then $\|f\|_\infty = e^{-1}$. $\|f\|_2 = (1/2)\sqrt{(1 - e^{-2a})/a}$.

(d) $1/\sqrt{1+x}$.
    Solution: $\|f\|_\infty = 1$, $\|f\|_2 = 0.9101797207$.

15. Define

$$f_n(x) = \begin{cases} 1 - nx, & 0 \leq x \leq \frac{1}{n}; \\ 0, & \frac{1}{n} \leq x \leq 1. \end{cases}$$

Show that

$$\lim_{n \to \infty} \|f_n(x)\|_\infty = 1,$$

but

$$\lim_{n \to \infty} \|f_n(x)\|_2 = 0.$$

Hint: Draw a graph of $f_n(x)$.

Solution: This can be done by direct computation but a graph of f_n is revealing. It shows that $\max |f_n| = 1$ for all n, but the support of $|f_n|$ goes to zero, thus forcing the integral norm to vanish while the pointwise norm remains at 1.

16. Show that

$$\|f\|_{1,[a,b]} = \int_a^b |f(x)| dx$$

defines a function norm.

Solution: We have, for any constant c,

$$\|cf\|_{1,[a,b]} = \int_a^b |cf(x)| dx = |c| \int_a^b |f(x)| dx = |c| \|f\|_{1,[a,b]},$$

and

$$
\begin{aligned}
\|f + g\|_{1,[a,b]} &= \int_a^b |f(x) + g(x)| dx \leq \int_a^b |f(x)| dx + \int_a^b |g(x)| dx \\
&= \|f\|_{1,[a,b]} + \|g\|_{1,[a,b]}.
\end{aligned}
$$

This establishes two of the three conditions. To establish the remaining condition requires an argument to the effect that the integral of the absolute value of a non-zero function must be greater than zero. Doing this precisely is probably beyond the analysis skills of most students, but good students ought to be able to construct a reasonable argument.

◁ • • ▷

4.4   APPLICATION: MULLER'S METHOD AND INVERSE QUADRATIC INTERPOLATION

Exercises:

1. Do three steps of Muller's method for $f(x) = 2 - e^x$, using the initial points $0, 1/2, 1$. Make sure you reproduce the values in the table in the text.

2. Repeat the above for inverse quadratic interpolation.

    Solution: We get $x_3 = 0.7087486790$, $x_4 = 0.692849951$, $x_5 = 0.692849951$.

3. Let $f(x) = x^6 - x - 1$; show that this has a root on the interval $[0, 2]$ and do three steps of Muller's method using the ends of the interval plus the midpoint as the initial values.

    Solution: $f(0) = -1$ and $f(2) = 61$, so a root exists on $[0, 2]$. We get $x_3 = 1.0313$, $x_4 = 1.1032$, $x_5 = 1.1258$.

4. Repeat the above using inverse quadratic interpolation.

   Solution: We get $x_3 = 1.017357687$, $x_4 = 1.032876088$, $x_5 = 1.174008963$.

5. Let $f(x) = e^x$ and consider the nodes $x_0 = -1$, $x_1 = 0$, and $x_2 = 1$. Let p_2 be the quadratic polynomial that interpolates f at these nodes, and let q_2 be the quadratic polynomial (in y) that *inverse* interpolates f at these nodes. Construct $p_2(x)$ and $q_2(y)$ and plot both, together with $f(x)$.

   Solution:
   $$p_2(x) = 0.543080635\, x^2 + 1.175201193\, x + 1.0,$$
   $$q_2(y) = -0.4254590638\, y^2 + 2.163953414\, y - 1.738494350.$$

6. Repeat the above using $f(x) = \sin \pi x$ and the nodes $x_0 = 0$, $x_1 = \frac{1}{4}$, and $x_2 = \frac{1}{2}$.

   Solution: We get $p_2(x) = -3.31370850x^2 + 3.656854248x$, $q_2(x) = y^2/2$.

7. Show that the formulas (4.15) and (4.16) for the divided-difference coefficients in Muller's method (4.14) are correct.

   Solution: Direct computation.

8. Show that the formulas (4.17) and (4.18) for the coefficients in inverse quadratic interpolation are correct.

   Solution: Direct computation.

9. Apply Muller's method and inverse quadratic interpolation to the functions in Exercise 3 of §3.1, and compare your results to those obtained with Newton's method and/or the secant method.

   Solution: See Table 4.8 for the Muller's method results.

Table 4.8   Solutions to Exercise 4.4.9

n	$f(x) = x^3 - 2x + 5$	$f(x) = e^x - 2$	$f(x) = x^2 - \sin x$
1	2.00000000000000	0.68725936775341	0.87590056539793
2	2.09020134845606	0.69308784769077	0.87670226627220
3	2.09451837842307	0.69314716126932	0.87672621454247
4	2.09455148276226	0.69314718055995	0.87672621539506
5	2.09455148154233	0.69314718055995	0.87672621539506

10. Refer back to the discussion of the error estimate for the secant method in §3.11.3. Adapt this argument to derive an error estimate for Muller's method.

    Solution:

    Muller's method is based on finding the exact root of a quadratic polynomial which interpolates the original function f. Let m be this interpolating polynomial, so that we have
    $$m(x) = f(x_n) + a_1(x - x_n) + a_2(x - x_n)(x - x_{n-1}),$$
    where a_1 and a_2 are divided difference coefficients and the next iterate is defined by solving $0 = m(x_{n+1})$. Then the quadratic interpolation error estimate implies that
    $$f(x) - m(x) = \frac{1}{6}(x - x_n)(x - x_{n-1})(x - x_{n-2})f'''(\xi_n)$$

for all x; here ξ_n is in the interval defined by x_n, x_{n-1}, x_{n-2}, and α. Set $x = \alpha$ so that we have

$$f(\alpha) - m(\alpha) = \frac{1}{6}(\alpha - x_n)(\alpha - x_{n-1})(\alpha - x_{n-2})f'''(\xi_n),$$

But $f(\alpha) = 0$ so we have

$$-m(\alpha) = \frac{1}{6}(\alpha - x_n)(\alpha - x_{n-1})(\alpha - x_{n-2})f'''(\xi_n).$$

On the other hand we know that $m(x_{n+1}) = 0$, so we can substitute this in to get

$$m(x_{n+1}) - m(\alpha) = \frac{1}{6}(\alpha - x_n)(\alpha - x_{n-1})(\alpha - x_{n-2})f'''(\xi_n).$$

The Mean Value Theorem then implies that there exists a value η_n between α and x_{n+1} such that $m(x_{n+1}) - m(\alpha) = m'(\eta_n)(x_{n+1} - \alpha)$; therefore,

$$m'(\eta_n)(x_{n+1} - \alpha) = \frac{1}{6}(\alpha - x_n)(\alpha - x_{n-1})(\alpha - x_{n-2})f'''(\xi_n)$$

or

$$x_{n+1} - \alpha = \frac{1}{6}(\alpha - x_n)(\alpha - x_{n-1})(\alpha - x_{n-2})\frac{f'''(\xi_n)}{m'(\eta_n)}.$$

To really tidy up the final result we need to relate $m'(\eta_n)$ to f'.

$$\triangleleft \bullet \bullet \bullet \triangleright$$

4.5   APPLICATION: MORE APPROXIMATIONS TO THE DERIVATIVE

Exercises:

1. Apply the derivative approximations (4.21) and (4.23) to the approximation of $f'(x)$ for $f(x) = x^3 + 1$ for $x = 1$ and $h = \frac{1}{8}$.

   Solution: Using both of (4.21) and (4.23) we get

   $$f'(x) \approx 2.968750000.$$

2. Apply the derivative approximations (4.21) and (4.23) to the same set of functions as in Exercise 3 of §2.2, using a decreasing sequence of mesh values, $h^{-1} = 2, 4, 8, \ldots$. Do we achieve the expected rate of accuracy as h decreases?

3. Derive a version of (4.24) under the assumption that $x_1 - x_0 = h$, but $x_2 - x_1 = \eta = \theta h$ for some real, positive, θ. Be sure to include the error estimate as part of your work, and confirm that when $\theta = 1$ you get the same results as in the text.

   Solution:

   $$f'(x_0) = \frac{-\theta(2 + \theta)f(x_0) + (1 + 2\theta + \theta^2)f(x_1) - f(x_2)}{\theta(1 + \theta)h} + \frac{1}{6}(1 + \theta)h^2 f'''(\xi).$$

4. Repeat the above for (4.25).

   Solution:

$$f'(x_1) = (-\eta/(h(h+\eta)))f(x_0) - ((h-\eta)/(h\eta))f(x_1) + (h/(\eta(h+\eta)))f(x_2)$$
$$- (h\eta/6)f'''(\xi).$$

5. Repeat the above for (4.26).

   Solution:

$$f'(x_2) = \frac{\theta^2 f(x_0) - (1+2\theta+\theta^2)f(x_1) + (2\theta+1)f(x_2)}{\theta(1+\theta)h} + \frac{1}{6}(1+\theta)h^2 f'''(\xi).$$

6. Use the derivative approximations from this section to construct a table of values for the derivative of the gamma function, based on the data in Table 4.3 (Table 4.8 in the text).

   Solution: $\Gamma'(1) \approx -0.5638283100$, $\Gamma'(1.5) \approx 0.3125765900 \times 10^{-1}$, $\Gamma'(2) \approx 0.4216022150$.

7. Try to extend the ideas of this section to construct an approximation to $f''(x_k)$. Is it possible? What happens?

   Solution: You can't eliminate all the terms that depend on the derivative with respect to x of ξ_x.

◁ ● ● ▷

4.6   HERMITE INTERPOLATION

Exercises:

1. Show that H_2, as defined in (4.28), is the cubic Hermite interpolate to f at the nodes $x = a$ and $x = b$.

   Solution: Clearly $H_2(a) = f(a)$ and $H_2'(a) = f'(a)$. In addition,

$$H_2(b) = f(a) + f'(a)(b-a) + (A - f'(a))(b-a) = f(a) + A(b-a) = f(b),$$

   and

$$H_2'(b) = f'(a) + 2B(b-a) + D(b-a)^2 = f'(a) + 2A - 2f'(a)$$
$$+ (C - B)(b-a)$$
$$= f'(a) + 2A - 2f'(a) + [(f'(b) - A) - (A - f'(a))]$$
$$= f'(a) - 2f'(a) + f'(a) + 2A - 2A + f'(b)$$
$$= f'(b).$$

   H_2 is a cubic polynomial and it reproduces the function and derivative values at the nodes. Therefore, by the uniqueness part of Theorem 4.4, H_2 must be the Hermite interpolant.

2. Construct the cubic Hermite interpolate to $f(x) = \sin x$ using the nodes $a = 0$ and $b = \pi$. Plot the error between the interpolate and the function.

   Solution: $H_2(x) = -(1/\pi)x^2 + x$.

3. Construct the cubic Hermite interpolate to $f(x) = \sqrt{1+x}$ using the nodes $a = 0$ and $b = 1$. Plot the error between the interpolate and the function.

   Solution:

   $$H_2(x) = 1.0 + 0.5000000000\,x - 0.085786438\,x^2 + 0.025126266\,x^2\,(x - 1.0).$$

   The maximum absolute error over $[0, 1]$ is about 0.0007.

4. Show that the error in cubic Hermite interpolation at the nodes $x = a$ and $x = b$ is given by

   $$\|f - H_2\|_\infty \leq \frac{(b-a)^4}{384}\|f^{(4)}\|_\infty.$$

   Solution: This is a direct application of Theorem 4.5, using $n = 2$.

5. Construct the cubic Hermite interpolate to $f(x) = \sqrt{x}$ on the interval $[\frac{1}{4}, 1]$. What is the maximum error on this interval, as predicted by theory? What is the maximum error that actually occurs (as determined by observation; no need to do a complete calculus max/min problem)?

   Solution:

   $$H_2(x) = 0.2037037037 + 1.388888889\,x - 0.8888888889\,x^2 + 0.2962962963\,x^3.$$

   The error theory predicts that the error is bounded according to

   $$
   \begin{aligned}
   |\sqrt{x} - H_2(x)| &\leq \frac{(3/4)^4}{384} \max_{x \in [1/4,1]} |(15/16)x^{-7/2}| = \frac{81}{256} \times \frac{1}{384} \times \frac{15}{16} \times 4^{7/2} \\
   &= 0.9887695312 \times 10^{-1}.
   \end{aligned}
   $$

   The actual error, looking at a computer plot, is about 0.0065.

6. Construct the cubic Hermite interpolate to $f(x) = 1/x$ on the interval $[\frac{1}{2}, 1]$. What is the maximum error as predicted by theory? What is the actual (observed) maximum error?

   Solution: $H_2(x) = 6 - 13x + 12x^2 - 4x^3$, the predicted error is $\leq 1/8$, the observed error is ≤ 0.022.

7. Construct the cubic Hermite interpolate to $f(x) = x^{1/3}$ on the interval $[\frac{1}{8}, 1]$. What is the maximum error as predicted by theory? What is the actual (observed) maximum error?

   Solution:

   $$H_2(x) = 0.3090379009 + 1.732750243\,x - 1.725947522\,x^2 + 0.6841593780\,x^3$$

   with maximum observed error about 0.036.

Table 4.9 Table for divided differences for quintic Hermite interpolation.

k	x_k	$f_0(x_k)$	$f_1(x_k)$	$f_2(x_k)$	$f_3(x_k)$	$f_4(x_k)$	$f_5(x_k)$
1	a	$f(a)$					
			$f'(a)$				
1	a	$f(a)$		B			
			A		D		
2	b	$f(b)$		C		F	
			$f'(b)$		E		G
2	b	$f(b)$		γ		θ	
			α		δ		
3	c	$f(c)$		β			
			$f'(c)$				
3	c	$f(c)$					

8. Construct the cubic Hermite interpolate to $f(x) = \ln x$ on the interval $[\frac{1}{2}, 1]$. What is the maximum error as predicted by theory? What is the actual (observed) maximum error?

   Solution: $H_2(x) = -2.227411278 + 4.364467670x - 3.046701500x^2 + .9096451100x^3$, the predicted error is $\leq 1/64$, the observed error is ≤ 0.0037.

9. Extend the divided difference table for cubic Hermite interpolation to *quintic* Hermite interpolation, using three nodes $x = a$, $x = b$, and $x = c$.

   Solution: We get a table like Table 4.9. Here A, B, C, and D are as in the text, and

$$\alpha = \frac{f(b) - f(c)}{b - c} \qquad \beta = \frac{\alpha - f'(c)}{b - c}$$
$$\gamma = \frac{f'(b) - \alpha}{b - c} \qquad \delta = \frac{\gamma - \beta}{b - c}$$
$$E = \frac{C - \gamma}{a - c} \qquad F = \frac{D - E}{a - c}$$
$$\theta = \frac{E - \delta}{a - c} \qquad G = \frac{F - \theta}{a - c}$$

10. Construct the quintic Hermite interpolate to $f(x) = \ln x$ on the interval $[\frac{1}{2}, 1]$; use $x = 3/4$ as the third node.

    Solution: We get

$$H_2(x) = -2.64516010 + 7.1200353\,x + 9.160601\,x^3 - 9.971884\,x^2 - 4.659891\,x^4$$
$$+ 0.9962986\,x^5.$$

Fig. 4.14 shows the error plot.

11. What is the error in quintic Hermite interpolation?

    Solution: We get something like

$$f(x) - H_3(x) = \frac{1}{720}\psi_3(x)f^{(6)}(\xi),$$

where $\psi_3(x) = (x - x_1)^2(x - x_2)^2(x - x_3)^2$. This can be bounded above according to

$$\|f - H_3\|_\infty \leq Ch^6\|f^{(6)}\|_\infty,$$

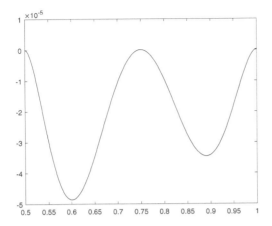

Figure 4.14   Error plot for Problem 10.

where $C = 0.0002057613167$, approximately. This value comes from using ordinary calculus to find the extreme values of ψ_3.

12. Extend the ideas of §4.5 to allow us to compute second derivative approximations using Hermite interpolation.

   Solution: Because $\psi_n(x)$ involves terms of the form $(x - x_k)^2$ for each node, the basic idea from §4.5 can be used to get derivative approximation formulas for the second derivative, based on Hermite interpolation.

4.7   PIECEWISE POLYNOMIAL INTERPOLATION

Exercises:

1. Use divided difference tables to construct the separate parts of the piecewise quadratic polynomial $q_2(x)$ that interpolates to $f(x) = \cos \frac{1}{2}\pi x$ at $x = 0, \frac{1}{4}, \frac{1}{2}, \frac{3}{4}, 1$. Plot the approximation and the error $\sin \frac{1}{2}\pi x - q_2(x)$.

2. Repeat the above using $f(x) = \sqrt{x}$ with the nodes $x = \frac{1}{5}, \frac{2}{5}, \frac{3}{5}, \frac{4}{5}, 1$.

   Solution: This requires that we construct the two divided difference tables that are required here. We have the results in Table 4.101, thus the piecewise polynomial is given by

   $$q_2(x) = \begin{cases} 0.4472 + 0.9262(x - 0.2) - 0.5388(x - 0.2)(x - 0.4) & 0.2 \le x \le 0.6 \\ 0.7746 + 0.5992(x - 0.6) - 0.1782(x - 0.6)(x - 0.8) & 0.6 \le x \le 1. \end{cases}$$

Table 4.10 Divided difference table.

k	x_k	$f_0(x_k)$	$f_1(x)$	$f_2(x)$
0	0.2	0.4472		
			0.9262	
1	0.4	0.6325		−0.5388
			0.7107	
2	0.6	0.7746		
k	x_k	$f_0(x_k)$	$f_1(x)$	$f_2(x)$
0	0.6	0.7746		
			0.5992	
1	0.8	0.8944		−0.1782
			0.5279	
2	1.0	1.0000		

3. Confirm that (4.30) is the correct piecewise quadratic approximation to $f(x) = 1/(1 + 25x^2)$ using the nodes $x_0 = -1$, $x_1 = -2/3$, $x_2 = -1/3$, $x_3 = 0$, $x_4 = 1/3$, $x_5 = 2/3$, and $x_6 = 1$.

Solution: Since each polynomial piece matches the nodal data it must be correct, because of the uniqueness of polynomial interpolation.

4. Using the data in Table 4.3 (Table 4.8 in the text), construct a piecewise cubic interpolating polynomial to the gamma function, using the nodes $x = 1.0, 1.2, 1.3, 1.5$ for one piece, and the nodes $x = 1.5, 1.7, 1.8, 2.0$ for the other piece. Use this approximation to estimate $\Gamma(x)$ for $x = 1.1, 1.4, 1.6$ and 1.9. How accurate is the approximation?

Solution:

$$q_3(x) = -0.3427614666x^3 + 1.873584557x^2$$
$$- 3.283390574x + 2.752567484, \quad \text{for } 1 \le x \le 3/2,$$
$$= (0.1596133333 \times 10^{-2})x^3 + 0.3766571466x^2$$
$$- 1.105518098x + 1.691638542, \quad \text{for } 3/2 \le x \le 2.$$

Fig. 4.15 shows the error plot. Note the obvious "kink" at $x = 1.5$.

5. Using the results of Exercise 6 from §4.6 of the text, together with the data of function values from Exercise 9 of §4.2, construct a piecewise cubic Hermite interpolating polynomial for the gamma function, using the nodes $x = 1.0, 1.3, 1.7, 2.0$. Test the accuracy of the interpolation by using it to approximate $\Gamma(x)$ for $x = 1.1, 1.2, 1.4, 1.5, 1.6, 1.8, 1.9$.

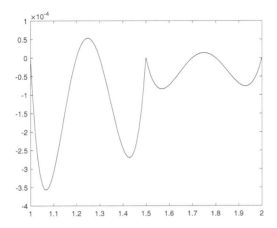

Figure 4.15   Error plot for Problem 4.

Solution: We get

$$q_3(x) = 2.663132928 - 3.038661703x + 1.651752931x^2$$
$$- 0.2762241567x^3, \quad \text{for } 1 \le x \le 1.3,$$
$$= 2.072404716 - 1.800684259x + 0.7983854147x^2$$
$$- (0.8343915950 \times 10^{-1})x^3, \quad \text{for } 1.3 \le x \le 1.7,$$
$$= 1.582234466 - 0.9241759694x + 0.2766990121x^2$$
$$+ (0.1991517800 \times 10^{-1})x^3, \quad \text{for } 1.7 \le x \le 2.$$

Moreover, we have

$$\Gamma(1.1) - q_3(1.1) = -0.2209795 \times 10^{-3}$$
$$\Gamma(1.2) - q_3(1.2) = 0.2209792 \times 10^{-3}$$
$$\Gamma(1.4) - q_3(1.4) = -0.612948 \times 10^{-4}$$
$$\Gamma(1.5) - q_3(1.5) = 0.885788 \times 10^{-4}$$
$$\Gamma(1.6) - q_3(1.6) = 0.1055826 \times 10^{-3}$$
$$\Gamma(1.8) - q_3(1.8) = 0.159327 \times 10^{-4}$$
$$\Gamma(1.9) - q_3(1.9) = -0.159317 \times 10^{-4}$$

6. Construct a piecewise cubic interpolating polynomial to $f(x) = \ln x$ on the interval $[\frac{1}{2}, 1]$, using the nodes

$$x_k = \frac{1}{2} + \frac{k}{18}, \quad 0 \le k \le 9.$$

Compute the value of the error $\ln x - p(x)$ at 500 equally spaced points on the interval $[\frac{1}{2}, 1]$, and plot the error. What is the maximum sampled error?

Solution: This is a fairly simple program to write, building upon previous modules. The maximum error which I got was $2.282428148858706 \times 10^{-5}$.

7. Repeat the above, using piecewise cubic Hermite interpolation over the same grid.

   Solution: The maximum error is about 1.9254×10^{-6}.

8. Construct piecewise polynomial approximations of degree $d = 1, 2, 3$ to the data in Table 4.7, using only the nodes $\log_{10} \theta_k = -6, -4, -2, 0, 2, 4, 6$. Plot the resulting curve and compare it to the ordinary interpolating polynomial found in Problem 13 of §4.2. Test how well this approximation matches the tabulated values at $\log_{10} \theta = -5, -3, -1, 1, 3, 5$.

9. Show that the error in piecewise cubic Hermite interpolation satisfies

$$\|f - H_2\|_\infty \le \frac{1}{384} h^4 \|f^{(4)}\|_\infty,$$

where we have assumed uniformly spaced nodes with $x_k - x_{k-1} = h$.

10. Given a grid of points

$$a = x_0 < x_1 < x_2 < \cdots < x_n = b$$

define the piecewise linear functions ϕ_k^h, $1 \le k \le n - 1$, according to

$$\phi_k^h(x) = \begin{cases} \frac{x - x_{k-1}}{x_k - x_{k-1}}, & x_{k-1} \le x \le x_k; \\ \frac{x_{k+1} - x}{x_{k+1} - x_k}, & x_k \le x \le x_{k+1}; \\ 0, & \text{otherwise.} \end{cases}$$

Define the function space

$$S_0^h = \{f \in C([a, b]), f(a) = f(b) = 0, f \text{ is piecewise linear on the given grid}\}.$$

Show that the ϕ_k^h are a basis for S_0^h, i.e., that every element of the space S_0^h can be written as a linear combination of the ϕ_k^h functions.

11. Implement a routine for approximating the natural logarithm using piecewise polynomial interpolation, i.e., a table look-up scheme. Assume that the table of (known) logarithm values is uniformly distributed on the interval $[\frac{1}{2}, 1]$. Choose enough points in the table to guarantee 10^{-10} accuracy for any x. Use

    (a) Piecewise linear interpolation;

    (b) Piecewise cubic Hermite interpolation.

Test your routine against the intrinsic logarithm function on your computer by evaluating the error in your approximation at $5,000$ equally spaced points on the interval $[\frac{1}{10}, 10]$. Use the way that the computer stores floating point numbers to reduce the logarithm computation to the interval $[\frac{1}{2}, 1]$, so long as $\ln 2$ is known to high accuracy.

◁ • • ▷

4.8   AN INTRODUCTION TO SPLINES

Exercises:

1. Given the set of nodes $x_0 = 0$, $x_1 = 1/2$, $x_2 = 1$, $x_3 = 3/2$, and $x_4 = 2$, we construct the cubic spline function

$$q_3(x) = 0.15B_{-1}(x) + 0.17B_0(x) + 0.18B_1(x) + 0.22B_2(x) + 0.30B_3(x)$$
$$+ 0.31B_4(x) + 0.32B_5(x),$$

where each B_k is computed from the exemplar B spline according to (4.38). Compute q_3 and its first derivative at each node.

Solution: We have

$$q_3(x_0) = 0.15B_{-1}(x_0) + 0.17B_0(x_0) + 0.18B_1(x_0) = 0.15 + 0.17 \times 4 + 0.18 = 1.01$$

and

$$q_3'(x_0) = 0.15B_{-1}'(x_0) + 0.17B_0'(x_0) + 0.18B_1'(x_0) = 0.15(-3/h) + 0.17(0)$$
$$+ 0.18(3/h) = 0.15(-6) + 0.18(6) = 0.18.$$

The others follow similarly.

2. Is the function

$$p(x) = \begin{cases} 0, & x \le 0; \\ x^2, & 0 \le x \le 1; \\ -2x^2 + 6x + 3, & 1 \le x \le 2; \\ (x-3)^2, & 2 \le x \le 3; \\ 0, & x \ge 3 \end{cases}$$

a spline function? Why/why not?

Solution: No, it is not, because it is not continuous at $x = 1$.

3. For what value of k is the following a spline function?

$$q(x) = \begin{cases} kx^2 + (3/2), & 0 \le x \le 1; \\ x^2 + x + (1/2), & 1 \le x \le 2. \end{cases}$$

Solution: There is no k that will make this a spline. Taking $k = 1$ makes the function continuous, but the derivative is not continuous.

4. For what value of k is the following a spline function?

$$q(x) = \begin{cases} x^3 - x^2 + kx + 1, & 0 \le x \le 1; \\ -x^3 + (k+2)x^2 - kx + 3, & 1 \le x \le 2. \end{cases}$$

Solution: $k = 3$.

5. For what value of k is the following a spline function?

$$q(x) = \begin{cases} x^3 + 3x^2 + 1, & -1 \le x \le 0; \\ -x^3 + kx^2 + 1, & 0 \le x \le 1. \end{cases}$$

Solution: $k = 3$.

6. Construct the natural cubic spline that interpolates to $f(x) = 1/x$ at the nodes $1/2, 5/8, 3/4, 7/8, 1$. Do this as a hand calculation. Plot your spline function and f on the same set of axes, and also plot the error.

    Solution: The coefficients are

    $$c = [0.4052, \ 0.3333, \ 0.2615, \ 0.2207, \ 0.1889, \ 0.1667, \ 0.1445].$$

7. Repeat the above using the complete spline.

    Solution: The coefficients are

    $$c = [0.4180, \ 0.3299, \ 0.2624, \ 0.2204, \ 0.1893, \ 0.1653, \ 0.1496].$$

8. Construct a natural spline interpolate to the mercury thermal conductivity data (Table 4.2, Table 4.7 in the text), using the 300° K, 500° K, and 700° K values. How well does this predict the values at 400° K and 600° K?

    Solution: The spline coefficients are $c_{-1} = 0.0095$, $c_0 = 0.0140$, $c_1 = 0.0185$, $c_2 = 0.0212$, $c_3 = 0.0239$; the spline predicts that the conductivity at 400° K $= 0.0972$ and at 600° K $= 0.1157$. Both values are accurate, but not exact.

9. Confirm that the function $B(x)$, defined in (4.35), is a cubic spline.

    Solution: It obviously is piecewise cubic; to confirm that it is a spline simply show that the one-sided limits at the interior knots are equal.

10. Construct a natural cubic spline to the gamma function, using the data in Table 4.3, and the nodes $x = 1.0, 1.2, 1.4, 1.6, 1.8$ and 2.0. Use this approximation to estimate $\Gamma(x)$ at $x = 1.1, 1.3, 1.5, 1.7$, and 1.9.

    Solution: Since this is a natural spline construction, we have the following system of equations to solve:

    $$\begin{bmatrix} 4 & 1 & 0 & 0 \\ 1 & 4 & 1 & 0 \\ 0 & 1 & 4 & 1 \\ 0 & 0 & 1 & 4 \end{bmatrix} \begin{bmatrix} c_0 \\ c_1 \\ c_2 \\ c_3 \end{bmatrix} = \begin{bmatrix} 0.7515020757 \\ 0.8872638175 \\ 0.8935153493 \\ 0.7647171043 \end{bmatrix}.$$

Thus

$$\begin{bmatrix} c_0 \\ c_1 \\ c_2 \\ c_3 \end{bmatrix} = \begin{bmatrix} 0.1511220252 \\ 0.1470139748 \\ 0.1480858929 \\ 0.1541578028 \end{bmatrix}.$$

The natural spline construction then tells us the rest of the coefficients:

$$c_0 = \Gamma(x_0)/6 = 1/6, \quad c_5 = \Gamma(x_5)/6 = 1/6;$$

and

$$c_{-1} = 2c_0 - c_1 = 0.18221130811994, \quad c_6 = 2c_5 - c_4 = 0.17917553048357.$$

Thus the spline is

$$q(x) = 0.1822113081B_{-1}(x) + 0.1666666667B_0(x) + 0.1511220252B_1(x)$$
$$+ 0.1470139748B_2(x) + 0.1480858929B_3(x) + 0.1541578028B_4(x)$$
$$+ 0.1666666667B_5(x) + 0.1791755304B_6(x).$$

We can test this by computing

$$t_k = c_{k-1} + 4c_k + c_{k+1};$$

we should get $t_k = f(x_k)$, and we do. Just for example,

$$t_2 = c_1+4c_2+c_3 = 0.1511220252+0.1470139748\times4+0.1480858929 = 0.8872638173,$$

which matches $\Gamma(x_2)$ to nine digits, and probably would match the tenth digit if I had been more careful with my rounding here. Similar results hold at the other points.

We can use this to approximate Γ at the indicated points by computing as follows:

$$q(1.1) = 0.1822113081B_{-1}(1.1) + 0.1666666667B_0(1.1) + 0.1511220252B_1(1.1)$$
$$+ 0.1470139748B_2(1.1),$$

$$q(1.3) = 0.1666666667B_0(1.3) + 0.1511220252B_1(1.3) + 0.1470139748B_2(1.3)$$
$$+ 0.1480858929B_3(1.3)$$

and so on for the other points.

11. Repeat the above using the complete spline approximation, and use the derivative approximations from §4.5 for the required derivative endpoint values.

    Solution: The spline coefficients are

    $$c = [0.1872,\ 0.1653,\ 0.1515,\ 0.1469,\ 0.1481,\ 0.1544,\ 0.1658,\ 0.1824].$$

    Moreover, $q_3(1.1) = 0.95256719297015$, which is a good approximation to $\Gamma(1.1) = 0.95135076986687$.

12. Repeat Problem 6 of §4.7, except this time construct the complete cubic spline interpolate to $f(x) = \ln x$, using the same set of nodes. Plot the approximation and the error. What is the maximum sampled error in this case?

    Solution: The spline coefficients are

    $$c = [-0.1345,\ -0.1153,\ -0.0977,\ -0.0819,\ -0.0674,\ -0.0541,\ -0.0417,$$
    $$-0.0303,\ -0.0195,\ -0.0094,\ 0.0001,\ 0.0090].$$

13. Recall Problem 7 from §4.2, in which we constructed polynomial interpolates to timing data from the 1973 Kentucky Derby, won by the horse Secretariat. For simplicity, we repeat the data in Table 4.11. Here t is the elapsed time (in seconds) since the race began and x is the distance (in miles) that Secretariat has travelled.

    (a) Construct a natural cubic spline that interpolates this data.

Table 4.11 Data for Problem 13.

x	0.0	0.25	0.50	0.75	1.00	1.25
t	0.0	25.0	49.4	73.0	96.4	119.4

(b) What is Secretariat's speed at each quarter-mile increment of the race? (Use miles per hour as your units.)

(c) What is Secretariat's "initial speed," according to this model? Does this make sense?

Note: It is possible to do this problem using a uniform grid. Construct the spline that interpolates t as a function of x, then use your knowledge of calculus to find $x'(t)$ from $t'(x)$.

Solution: (a) Again, the fact that we are asked to do a natural spline tells us the linear system is

$$
\begin{bmatrix}
4 & 1 & 0 & 0 \\
1 & 4 & 1 & 0 \\
0 & 1 & 4 & 1 \\
0 & 0 & 1 & 4
\end{bmatrix}
\begin{bmatrix}
c_1 \\ c_2 \\ c_3 \\ c_4
\end{bmatrix}
=
\begin{bmatrix}
25.0 \\ 49.4 \\ 73.0 \\ 76.5
\end{bmatrix}.
$$

Thus, the first four spline coefficients are

$$
\begin{bmatrix}
c_1 \\ c_2 \\ c_3 \\ c_4
\end{bmatrix}
=
\begin{bmatrix}
4.1842105263 \\
8.2631578947 \\
12.1631578947 \\
16.0842105263
\end{bmatrix},
$$

and the remaining coefficients are

$$
c_0 = 0, \quad c_{-1} = -4.1842105263, \quad c_5 = 19.9, \quad c_6 = 23.71578947368422.
$$

(b) Our spline, as constructed, gives us the time, t (in seconds) as a function of the distance, x (in miles). But velocity is the derivative of distance with respect to time. We use the calculus to tell us that

$$
\frac{dx}{dt} = \left(\frac{dt}{dx} \right)^{-1}.
$$

Thus we can compute $q'(1.25)$ and from that get the horse's speed. We get

$$
q'(1.25) = (-3/h)(16.0842105263) + (3/h)(23.7157894736),
$$

so

$$
\frac{dx}{dt} = 0.0109195402.
$$

This seems a little low, but then we remember that the units are such that this is *miles per second*. Converting to *miles per hour* we get

$$
\frac{dx}{dt} = 39.3103448279,
$$

which is a much more reasonable speed.

(c) The same kind of computation gives us an initial speed of 35.84905660390886. I suspect the horse took a few seconds to get up to full speed, so this is probably a bit high.

14. Show that the complete cubic spline problem can be solved (approximately) without having an explicit expression for f' at the endpoints. Hint: Consider the material from §4.5.

    Solution: Simply use the approximations

    $$f'(b) \approx \frac{3f(b) - 4f(b-h) + f(b-2h)}{2h}$$

    and

    $$f'(a) \approx \frac{-f(a+2h) + 4f(a+h) - 3f(a)}{2h}.$$

15. Construct the natural cubic spline that interpolates the data in Table 4.7 (Table 4.12 in the text) at the nodes defined by $\log_{10} \theta = -6, -4, \ldots, 4, 6$. Test the accuracy of the approximation by computing $q_3(x)$ for $x = \log_{10} \theta = -5, -3, \ldots, 3, 5$ and comparing to the actual data in the table.

    Solution: The spline coefficients are $c_{-1} = -0.0029$, $c_0 = 0.0002$, $c_1 = 0.0032$, $c_2 = -0.0037$, $c_3 = 0.1056$, $c_4 = 0.3492$, $c_5 = 0.4793$, $c_6 = 0.5904$, $c_7 = 0.7016$.

16. Construct the exemplar quadratic B-spline; i.e., construct a piecewise quadratic function that is C^1 over the nodes/knots $x = -1, 0, 1, 2$, and which vanishes for x outside the interval $[-1, 2]$.

    Solution: A quadratic spline is a piecewise quadratic function with first derivative continuity at the knots; to be the exemplar B-spline we also need the local definition property, thus we need our function to be zero for $x \leq -1$ and $x \geq 2$. This leads us to try something like

    $$B(x) = \begin{cases} 0, & x \leq -1; \\ (x+1)^2, & -1 \leq x \leq 0; \\ Ax^2 + Bx + C; & 0 \leq x \leq 1; \\ (x-2)^2; & 1 \leq x \leq 2; \\ 0, & 2 \leq x. \end{cases}$$

    We need to choose A, B, and C so that $B \in C^1$, i.e., so that

    $$\lim_{x \to 0^-} B(x) = \lim_{x \to 0^+} B(x),$$

    $$\lim_{x \to 1^-} B(x) = \lim_{x \to 1^+} B(x),$$

    and similarly for B'. This quickly leads, after a modest amount of manipulation, to the values

    $$C = 1, \quad B = 2, \quad A = -2.$$

So the exemplar quadratic B-spline is

$$B(x) = \begin{cases} 0, & x \le -1; \\ (x+1)^2, & -1 \le x \le 0; \\ -2x^2 + 2x + 1; & 0 \le x \le 1; \\ (x-2)^2; & 1 \le x \le 2; \\ 0, & 2 \le x. \end{cases}$$

17. Construct the exemplar *quintic* B-spline.

    Solution: This is lengthy but not hard. Based on the cubic construction, we are going to need to use nodes at $x = -3$, $x = -2$, $x = -1$, $x = 0$, $x = 1$, $x = 2$, and $x = 3$. In order to get the proper derivative continuity at the nodes $x = \pm 3$ we will need the exemplar quintic to have the basic form

$$B(x) = \begin{cases} 0, & x < -3 \\ (x+3)^5, & -3 < x \le -2, \\ q_1(x), & -2 < x \le -1, \\ q_2(x), & -1 < x \le 0, \\ q_3(x), & 0 < x \le 1, \\ q_4(x), & 1 < x \le 2, \\ (x-3)^5, & 2 < x \le 3, \\ 0, & 3 < x. \end{cases}$$

Here we have to choose the individual quintics q_1, q_2, etc., to maintain continuity and smoothness. For example, we have to have that

$$q_1(-2) = 1, \quad q_1'(-2) = 5, \quad q_1''(-2) = 20, \quad q_1'''(-2) = 60.$$

Therefore, we have to have

$$q_1(x) = 1 + 5(x+2) + 10(x+2)^2 + 10(x+2)^3 + A_1(x+2)^4 + B_1(x+2)^5.$$

Similarly, we can get that

$$q_4(x) = -1 + 5(x-2) - 10(x-2)^2 + 10(x-2)^3 + A_4(x-2)^4 + B_4(x-2)^5.$$

Corresponding relationships hold for q_2 and q_3. Algebraic equations for A_1, B_1, etc., can then be derived by setting $q_1(-1) = q_2(-1)$ and so forth.

18. For a linear spline function we have $d = 1$ which forces $N = 0$. Thus, a linear spline has no derivative continuity, only function continuity, and no additional conditions are required at the endpoints. Show that the exemplar B-spline of first degree is given by

$$B(x) = \begin{cases} 0, & x \le -1; \\ x+1, & -1 \le x \le 0; \\ 1-x, & 0 \le x \le 1; \\ 0, & 1 \le x. \end{cases}$$

Write out, in your own words, how to construct and evaluate a linear spline interpolant using this function as the basic B-spline.

Solution: What makes this spline basis so appealing is that the coefficients can be shown to be the nodal values. That is, we construct the spline approximation to a function f as

$$q_1(x) = \sum_{i=0}^{n} f(x_i) B_i(x).$$

19. Discuss, in your own words, the advantages or disadvantages of spline approximation compared to ordinary piecewise polynomial approximation.

    Solution: The main difference is the smoothness of the spline approximation, compared to the ordinary piecewise polynomial one.

20. Write an essay which compares and contrasts piecewise cubic Hermite interpolation with cubic spline interpolation.

    Solution: The main point of comparison would be that the Hermite process requires derivative values.

21. The data below gives the actual thermal conductivity data for the element nickel. Construct a natural spline interpolate to this data, using only the data at $200°$ K, $400°$ K, ..., $1400°$ K. How well does this spline predict the values at $300°$ K, $500°$ K, etc.?

Table 4.12   Data for Problem 21.

Temperature (K), u	200	300	400	500	600
Conductivity (watts/cm K), k	1.06	0.94	0.905	0.801	0.721
Temperature (K), u		700	800	900	1,000
Conductivity (watts/cm K), k		0.655	0.653	0.674	0.696

Solution: The spline coefficients are $c_{-1} = 0.2002$, $c_0 = 0.1767$, $c_1 = 0.1532$, $c_2 = 0.1156$, $c_3 = 0.1053$, $c_4 = 0.1160$, $c_5 = 0.1267$.

22. Construct a natural spline interpolate to the thermal conductivity data in the table below. Plot the spline and the nodal data.

Temperature (° K), u	200	300	400	500	600	700
Conductivity (watts/cm ° K), k	0.94	0.803	0.694	0.613	0.547	0.487

◁ • • ▷

4.9   TENSION SPLINES

Exercises:

1. Show that the piecewise function defined in (4.56) is, indeed, continuous and has continuous first and second derivatives.

Solution: This is a straight-forward, if tedious, calculus exercise. Simply show that the function values plus the first two derivatives are continuous at the knots ± 2, ± 1, 0, i.e., the limits from the left and right are equal. This is a good exercise to use something like Maple or Mathematica on.

2. Fill in the details of the natural spline construction. In particular, confirm the expressions for $\delta_{2,0}(p)$ and $\delta_{2,1}(p)$ (and that $\tau''(1) = \tau''(-1)$) as well as the form of the final linear system (4.59).

3. Derive the linear system for construction of a complete taut spline, by following what was done in §4.8.

4. Consider the following dataset:

x	600	650	700	750	800	850	900	950	1,000	1,050	1,100
y	0.64	0.65	0.66	0.69	0.91	2.2	1.2	0.62	0.6	0.61	0.61

Plot the data, and construct a (natural) polynomial spline fit to it. Note the "wiggles" to the left of the peak, which appear to be contrary to the sense of the data, which is increasing monotonically towards the peak near $x = 850$. Find the smallest value of p in a taut natural spline fit to this data which yields a monotone curve to the left of the peak.

Solution: The plots below show the polynomial spline (on the left) and a tension spline with $p = 4$ (on the right. Note that the wiggles have been smoothed out.

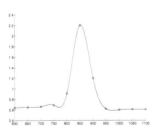

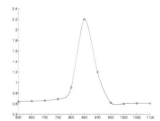

Figure 4.16   Polynomial (natural) spline fit to the data in Problem 4.

Figure 4.17   Tension spline fit ($p = 4$) to the data in Problem 4.

5. Repeat the previous problem using complete splines. Use a simple finite difference approximation based on the data to get the necessary derivative values.

6. Use pchip to plot the data in Fig. 4.35 (in the text) and compare this to the results you get with a tension spline.

7. Repeat the above, using spline.

◁ • • • ▷

4.10   LEAST SQUARES CONCEPTS IN APPROXIMATION

Exercises:

1. Modify the methods of §4.11.1 to compute the linear function of two variables that gives the best least-squares fit to the data in Table 4.14.

   Solution: We are looking for a linear fit in two variables so the model equation is $z = ax + by + c$. Then we want to minimize the function

   $$F(a, b, c) = \sum_{k=1}^{n} (z_k - (ax_k + by_k + c))^2.$$

   The critical point is defined by the three equations

   $$\frac{dF}{da} = -2 \sum_{k=1}^{n} (z_k - (ax_k + by_k + c))(-x_k) = 0,$$

   $$\frac{dF}{db} = -2 \sum_{k=1}^{n} (z_k - (ax_k + by_k + c))(-y_k) = 0,$$

   $$\frac{dF}{dc} = -2 \sum_{k=1}^{n} (z_k - (ax_k + by_k + c))(1) = 0,$$

   from which we get the 3×3 system

   $$\sum_{k=1}^{n} z_k x_k = \left(\sum_{k=1}^{n} x_k^2 \right) a + \left(\sum_{k=1}^{n} x_k y_k \right) b + \left(\sum_{k=1}^{n} x_k \right) c,$$

   $$\sum_{k=1}^{n} z_k y_k = \left(\sum_{k=1}^{n} x_k y_k \right) a + \left(\sum_{k=1}^{n} y_k^2 \right) b + \left(\sum_{k=1}^{n} y_k \right) c,$$

   $$\sum_{k=1}^{n} z_k = \left(\sum_{k=1}^{n} x_k \right) a + \left(\sum_{k=1}^{n} y_k \right) b + \left(\sum_{k=1}^{n} 1 \right) c,$$

   which can be solved to produce the coefficients.

2. The data in Table 4.13 (Table 4.27 in the text) gives the actual thermal conductivity data for the element iron. Construct a quadratic least squares fit to this data and plot both the curve and the raw data. How well does your curve represent the data? Is the fit improved any by using a cubic polynomial?

   Solution: $p_2(T) = 1.3873 - 0.00186222727273\,T + 0.00000088863636\,T^2$, and $p_3(T) = 1.5755 - 0.00353175602176\,T + 0.00000450850816\,T^2 - 0.00000000219386\,T^3$. While the two curves are close together, the cubic one has a concavity change that

Table 4.13   Data for Problem 2.

Temperature (K), u	100	200	300	400	500
Conductivity (watts/cm K), k	1.32	0.94	0.835	0.803	0.694
Temperature (K), u	600	700	800	900	1000
Conductivity (watts/cm K), k	0.613	0.547	0.487	0.433	0.38

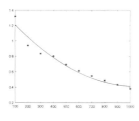   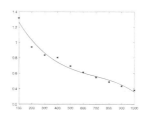

Figure 4.18   **Figure 4.19**

appears to be present in the raw data. Figs. 4.18–4.19 plot the curves, with the raw data given by asterisks.

3. Repeat Exercise 2, this time using the data for nickel from Exercise 21 in §4.8.

   Solution: The quadratic polynomial is

   $$p_2(T) = 1.36016190476191 - 0.00163075324675T + 0.00000095562771T^2$$

   and the cubic polynomial is

   $$p_3(T) = 1.19442857142860 - 0.00053271284271T - 0.00000109891775T^2$$
   $$+ 0.00000000114141T^3.$$

   Figure 4.20 shows both curves and the actual data.

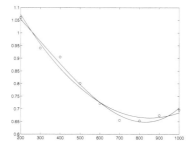

Figure 4.20   Solution plot for Exercise 4.10.3.

4. Modify the methods of §4.10.1 to compute the quadratic polynomial that gives the best least-squares fit to the data in Table 4.14.

Solution: The general quadratic is $y = ax^2 + bx + c$, so the least squares function that we want to minimize is

$$F(a, b, c) = \sum_{k=1}^{n}(y_k - (ax_k^2 + bx_k + c))^2.$$

The critical point is defined by the three equations

$$\frac{dF}{da} = -2\sum_{k=1}^{n}(y_k - (ax_k^2 + bx_k + c))(-x_k^2) = 0,$$

$$\frac{dF}{db} = -2\sum_{k=1}^{n}(y_k - (ax_k^2 + bx_k + c))(-x_k) = 0,$$

$$\frac{dF}{dc} = -2\sum_{k=1}^{n}(y_k - (ax_k^2 + bx_k + c))(1) = 0,$$

which simplifies to the 3×3 system

$$\sum_{k=1}^{n} y_k x_k^2 = \left(\sum_{k=1}^{n} x_k^4\right)a + \left(\sum_{k=1}^{n} x_k^3\right)b + \left(\sum_{k=1}^{n} x_k^2\right)c,$$

$$\sum_{k=1}^{n} y_k x_k = \left(\sum_{k=1}^{n} x_k^3\right)a + \left(\sum_{k=1}^{n} x_k^2\right)b + \left(\sum_{k=1}^{n} x_k\right)c,$$

$$\sum_{k=1}^{n} y_k = \left(\sum_{k=1}^{n} x_k^2\right)a + \left(\sum_{k=1}^{n} x_k\right)b + \left(\sum_{k=1}^{n} 1\right)c,$$

which can then be solved to get the coefficients.

Table 4.14   Data for Problem 4.

Problem 4		Problem 1		
x_n	y_n	x_n	y_n	z_n
−1	0.9747	0	0	0.9573
0	0.0483	0	1	2.0132
1	1.0223	1	0	2.0385
2	4.0253	1	1	1.9773
3	9.0152	0.5	0.5	1.9936

5. The data in Table 4.15 (Table 4.29 in the text) are supposed to represent a curve of the form $M = CN^r$, for some r. By taking logarithms of both sides and then doing a linear least squares fit, estimate the value of r. Plot your curve and the data.

6. The data in Table 4.16 (Table 4.30 in the text) are supposed to represent a relation of the form $E = CN^{-r}$, for some r. By taking logarithms, set this up as a linear least-squares problem and find an estimate for r. Plot the data and resulting curve.

Table 4.15 Data for Problem 5.

N	8	12	16	24	32	48
M	66	172	335	850	1,631	4,050

Table 4.16 Data for Problem 6.

N	8	12	16	24	32	48
E	5.24×10^{-5}	1.37×10^{-5}	4.73×10^{-6}	1.84×10^{-6}	9.21×10^{-7}	3.83×10^{-7}

7. An astronomical tracking station records data on the position of a newly discovered asteroid orbiting the Sun. The data are reduced to measurements of the radial distance from the Sun (measured in millions of kilometers) and angular position around the orbit (measured in radians), based on knowledge of the Earth's position relative to the Sun. In theory, these values should fit into the polar coordinate equation of an ellipse, given by

$$r = \frac{L}{2(1 + \epsilon \cos \theta)},$$

where ϵ is the eccentricity of the elliptical orbit and L is the width of the ellipse (sometimes known as the *latus rectum* of the ellipse) at the focus. (See Figure 4.21.) However, errors in the tracking process and approximations in the transformation to (r, θ) values perturb the data. For the data in Table 4.17, find the eccentricity of the orbit by doing a least squares fit to the data. Hint: Write the polar equation of the ellipse as

$$2r(1 + \epsilon \cos \theta) = L,$$

which can then be written as

$$2r = \epsilon(-2r \cos \theta) + L.$$

So, $y_k = 2r_k$ and $x_k = -2r_k \cos \theta_k$.

Table 4.17 Data for Problem 7.

r_n	42,895	42,911	42,851	42,779	42,774	42,764	42,750	42,744	42,749	42,749	42,894
θ_n	−0.1289	−0.1352	−0.1088	−0.0632	−0.0587	−0.0484	−0.0280	−0.0085	0.0259	0.0264	0.1282

Solution: $b = L = 149,670$, $m = \epsilon = 0.750825$, so the polar equation of the orbit is

$$r = \frac{149670}{2(1 + 0.750825 \cos \theta)}.$$

8. Prove Theorem 4.8.

Solution: All three required properties follow directly from the hypotheses, the properties of the definite integral, and the definition of $(\cdot, \cdot)_w$.

9. Let $w(x) = x$ on the interval $[0, 1]$; compute $\|f\|_w$ for each of the following functions:

   (a) e^{-x};

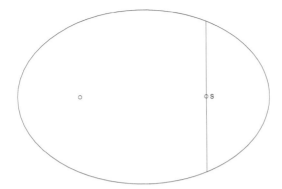

Figure 4.21   Figure for Problem 7. The closed curve is the elliptical orbit, and the vertical line has length L.

   Solution: $\|f\|_w = (1/2)\sqrt{1 - 3e^{-2}} = 0.3853550799,\ \|f\|_2 = 0.6575198540.$

(b) $1/\sqrt{x^2 + 1}$;
   Solution: $\|f\|_w = 0.5887050110,\ \|f\|_2 = 0.8862269255.$

(c) $1/\sqrt{x}$.
   Solution: $\|f\|_w = 1,\ \|f\|_2 =$ undefined.

Compare to the values obtained using the unweighted 2-norm.

10. Derive the linear system (4.61) as the solution to the least squares approximation problem.

   Solution: Take the derivative of R_n with respect to each C_k.

11. Provide the missing details to show that the family of polynomials defined in (4.63) is, indeed, orthogonal.

   Solution: Directly compute the inner product (ϕ_k, ϕ_i) for some $i < k$.

12. Let $\{\phi_k\}$ be a family of orthogonal polynomials associated with a general weight function w and an interval $[a, b]$. Show that the $\{\phi_k\}$ are independent in the sense that

$$0 = c_1 \phi_1(x) + c_2 \phi_2(x) + \cdots + c_n \phi_n(x)$$

   holds for all $x \in [a, b]$ if and only if $c_k = 0$ for all k.

   Solution: Take the inner product of the given relation with each ϕ_k. The result is the equation

$$0 = c_k (\phi_k, \phi_k) \Longrightarrow c_k = 0.$$

13. Prove the expansion formula (4.64) for a polynomial.

   Solution: This is really an exercise in linear algebra more than anything else. Since the $\{\phi_k\}$ are a basis for polynomials, it follows that we can write any polynomial as a linear combination of the ϕ_k. Thus

$$q_k = \sum_{i=0}^{k} C_k \phi_k$$

for some coefficients C_k. The precise form of the coefficient comes from using the orthogonality of the polynomials.

14. Construct the second degree Legendre least-squares approximation to $f(x) = \cos \pi x$ over the interval $[-1, 1]$.

    Solution: $p_2(x) = -2.279726631x^2 + 0.7599088770$; this is not especially accurate.

15. Construct the second degree Laguerre least-squares approximation to $f(x) = e^x$ on the interval $[0, \infty)$.

16. Construct the third degree Legendre least-squares approximation to $f(x) = \sin \frac{1}{2} \pi x$ over the interval $[-1, 1]$.

    Solution:
    $$p_3(x) = 60 \frac{x \left(-4 \pi^2 + 7 x^2 \pi^2 - 70 x^2 + 42\right)}{\pi^4}.$$

◁ • • ▷

4.11 ADVANCED TOPICS IN INTERPOLATION ERROR

Exercises:

1. Use your computer's random number function to generate a set of random values r_k, $0 \le k \le 8$, with $-0.1 \le r_k \le 0.1$. Construct the interpolating polynomial to $f(x) = \sin \pi x$ on the interval $[-1, 1]$, using 9 equally spaced nodes; call this $p_8(x)$. Then construct the polynomial that interpolates the same function at the same nodes, *except* perturb the function values using the r_k values; call this $\hat{p}_8(x)$. How much difference is there between the divided difference coefficients for the two polynomials? Plot both p_8 and $\hat{p}_8$ and comment on your results. Note: It is important here that you look at x values *between* the nodes. Do not produce your plots based simply on the values at the nodes. Repeat using a 16 degree interpolation and 16 random perturbation values.

    Solution: The precise results will depend on the random perturbations, but it should happen that the interpolant to the perturbed data is non-trivially different from the interpolant to the original data, especially for the higher-degree case. A script that the author used is given here, and plots for an 8 degree case and a 16 degree case are given below the script (the original interpolant is the solid line, the randomly perturbed interpolant is the dashed-dot line).

```
%
xc = [-8:8]/16;
yc = sin(pi*xc);
r = rand(17,1) - 0.5;
rk = r/10;
yr = yc + rk';
%
a = divdif(xc,yc);    % divdif defined earlier in this chapter
ar = divdif(xc,yr);
%
xx = [-100:100]/200;
yy = intval(xx,xc,a); % intval defined earlier in this chapter
yr = intval(xx,xc,ar);
%
figure(1)
plot(xx,yy,'k-',xx,yr,'k-.')
figure(2)
plot(xx,yy-yr,'k-')
%
```

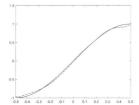

Figure 4.22

Figure 4.23

2. Repeat the above, using $f(x) = e^x$.

Solution: The precise results will again depend on the random perturbations, but it should again happen that the interpolant to the perturbed data is different from the interpolant to the original data, especially for the higher degree case.

Using an obvious modification of the above script, the author got the plots below.

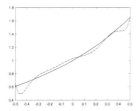

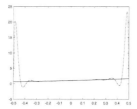

Figure 4.24

Figure 4.25

The point of both problems is that random small perturbations of the interpolation data can yield non-trivial changes to the resulting polynomial.

3. Use the Newton interpolating algorithm to construct interpolating polynomials of degree 4, 8, 16, and 32 using equally spaced nodes on the interval $[-1, 1]$, to the function $g(x) = (1 + 4x^2)^{-1}$. Is the sequence of interpolates converging to g throughout the interval?

   Solution: I got that $\|g - p_n\|_\infty$ went from 0.1618 (for $n = 4$) to 0.0671 for $n = 32$. The sequence of polynomials appears to be converging.

4. Use the Newton interpolating algorithm to construct interpolating polynomials of degree 4, 8, 16, and 32 using equally spaced nodes on the interval $[-1, 1]$, to the function $g(x) = (1 + 100x^2)^{-1}$. Is the sequence of interpolates converging to g throughout the interval?

   Solution: I got that $\|g - p_n\|_\infty$ went from 0.6149 (for $n = 4$) to 2.5×10^5 for $n = 32$. The sequence of polynomials appears to be diverging.

5. Write up a complete proof of Theorem 4.10, providing all the details left out in the text.

   Solution: This is straightforward. Simply go through the induction argument to establish the three-term recurrence, from which everything else follows.

6. Construct interpolating polynomials of degree 4, 8, 16, and 32 on the interval $[-1, 1]$ to $f(x) = e^x$ using equidistant and Chebyshev nodes. Sample the respective errors at 500 equally spaced points and compare your results.

   Solution: The Lagrange interpolation errors go like

   $$e_4 = 1.124354920726489 \times 10^{-3}, \quad e_8 = 5.799995905775290 \times 10^{-8},$$

   $$e_{16} = 1.532107773982716 \times 10^{-14}, \quad e_{32} = 3.647070978551880 \times 10^{-10},$$

   and the Chebyshev interpolation errors go like.

   $$e_4 = 6.396994825514923 \times 10^{-4}, \quad e_8 = 1.219007073061107 \times 10^{-8},$$

   $$e_{16} = 8.881784197001252 \times 10^{-16}, \quad e_{32} = 8.881784197001252 \times 10^{-16}.$$

   Clearly the Chebyshev errors are smaller.

7. The Chebyshev nodes are defined on the interval $[-1, 1]$. Show that the change of variable

   $$t_k = a + \frac{1}{2}(b - a)(x_k + 1)$$

   will map the Chebyshev nodes to the interval $[a, b]$.

   Solution: Since x_k are the Chebyshev nodes, they are cosine values and hence $|x_k| \le 1$. Therefore

   $$t_k \le a + \frac{1}{2}(b - a)(2) = b,$$

   and

   $$t_k \ge a + \frac{1}{2}(b - a)(0) = a.$$

Thus $t_k \in [a, b]$. Note that we could have used any transformation that carried $[-1, 1]$ into $[a, b]$, but obviously we want to keep things simple, which is why we chose the linear transformation that did the job.

8. Use the above to find the Chebyshev nodes for linear interpolation on the interval $[\frac{1}{4}, 1]$. Use them to construct a linear interpolate to $f(x) = \sqrt{x}$. Plot the error in this interpolation. How does it compare to the error estimate used in §3.7?

   Solution: The original Chebyshev nodes for linear interpolation are

   $$x_0 = \frac{1}{2}\sqrt{2}$$

   and

   $$x_1 = -\frac{1}{2}\sqrt{2}.$$

   The transformed values for $[1/4, 1]$ become

   $$t_0 = \frac{1}{4} + \frac{1}{2} \times \frac{3}{4} \times \frac{2 + \sqrt{2}}{2} = 0.890165043$$

   and

   $$t_1 = \frac{1}{4} + \frac{1}{2} \times \frac{3}{4} \times \frac{2 - \sqrt{2}}{2} = 0.359834957.$$

   The linear interpolate is then

   $$p_1(x) = 1.7790534738 \times (x - 0.359834957) + 1.131111480 \times (0.890165043 - x).$$

9. Repeat the above for $f(x) = x^{-1}$ on the interval $[\frac{1}{2}, 1]$.

   Solution: The Chebyshev nodes for linear interpolation are

   $$x_0 = \cos(\pi/4) = \frac{\sqrt{2}}{2},$$

   and

   $$x_1 = \cos(3\pi/4) = -\frac{\sqrt{2}}{2}.$$

   Applying the transformation into the interval $[\frac{1}{2}, 1]$ results in the new values

   $$t_0 = \frac{1}{2} + \frac{1}{2} \times \frac{1}{2} \times (x_0 + 1) = \frac{1}{2} + \frac{1}{4}\left(\frac{2 + \sqrt{2}}{2}\right) = 0.9267766953,$$

   and

   $$t_1 = \frac{1}{2} + \frac{1}{2} \times \frac{1}{2} \times (x_1 + 1) = \frac{1}{2} + \frac{1}{4}\left(\frac{2 - \sqrt{2}}{2}\right) = 0.5732233047,$$

   so the interpolate is easily computed as

   $$
   \begin{aligned}
   p_1(x) &= \frac{x - t_0}{t_1 - t_0}(1/t_1) + \frac{t_1 - x}{t_1 - t_0}(1/t_0) \\
   &= 3.051897118 \times (x - 0.9267766953) + 4.934250059 \times (0.5732233047 - x).
   \end{aligned}
   $$

This gives a much better approximation than the simple interpolate using the endpoints that we used in §3.4.

10. What is the error estimate for Chebyshev interpolation on a general interval $[a, b]$? In other words, how does the change of variable impact the error estimate?

    Solution: Let $f(x)$ be the function on $[a, b]$ that we wish to interpolate. To do this with the Chebyshev nodes, we construct the new function

    $$F(z) = f(\xi(z)),$$

    where $\xi(z) = a + (b - a)(z + 1)/2$. Then the interpolation error is controlled by the derivative (with respect to z) $F^{(n+1)}$. But the chain rule says that

    $$F^{(n+1)}(z) = f^{(n+1)}(\xi(z))[(b - a)/2]^{n+1}.$$

    Therefore, the error goes like

    $$\|f - p_n\|_\infty \leq \frac{(b - a)^{n+1}}{4^n(n + 1)!}\|f^{(n+1)}\|_\infty.$$

    The factor of 2^{-n} is replaced by

    $$\frac{(b - a)^{n+1}}{4^n}.$$

11. Let $f(x) = e^{-x^2}$; take $N = 2$ and construct both the Lobatto and Chebyshev node spectral interpolants. This is straight-forward enough that it can be done as a pencil-and-paper calculation, using just a few digits, but feel free to use whatever computing resources you wish. Show that the interpolating condition ($y_k = f(x_k)$) is satisfied. Repeat this for $N = 3$, and comment on your results compared to those for $N = 2$.

    Solution: Using the Lobatto nodes, for $N = 2$ we get

    $$b_i = (1.3679, -0.0, -0.6321),$$

    for $0 \leq i \leq 2$, which leads to the approximation and error plots in Figs. 4.26 and 4.27, with a maximum error of 0.0779.

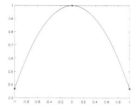

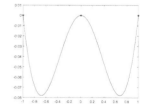

Figure 4.26   **Figure 4.27**

Using the Chebyshev nodes, for $N = 2$ we get

$$b_i = (1.3679. - 0.0, -0.6321),$$

for $0 \leq i \leq 2$, which leads to the approximation and error plots in Figs. 4.28 and 4.29,

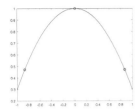

Figure 4.28   **Figure 4.29**

with a maximum error of 0.0714.

What is curious here is that the $N = 2$ interpolation (above) is only slightly less accurate than the $N = 3$ interpolation. This is largely because the function being approximated (e^{-x^2}) is an *even* function, so the addition of the odd-degree polynomials really don't do much. If we go to $N = 4$, we get that (for both sets of nodes) the error is $\mathcal{O}(0.05)$, and for $N = 6$, we have that it is $\mathcal{O}(4 \times 10^{-4})$.

12. Repeat the above for $f(x) = xe^{-x}$.

13. Write a script to do spectral interpolation (both sets of nodes); confirm that it reproduces the results in the previous exercises and examples, then use it to compute interpolants to the following examples:

    (a) $y = \sqrt{1 + x}$;

    Solution: Taking $N = 40$ we have that the maximum error is still 0.0087 (for the Chebyshev nodes). The problem is the singularity in the function as x approaches -1.

    (b) $y = \log 1.01 + x$.

    Solution: Similar results, again because of the singularity *near* $x = -1$.

    What value of N is needed to get accuracy of 10^{-5}? What about 10^{-12}? What seems to be the problem?

14. Let $f(x) = |x|^{2/3}$. Using either set of nodes, try to get a spectral interpolant that is accurate to within 10^{-5} over all of $[-1, 1]$. How many terms do you have to take? What seems to be the problem?

    Solution: Taking $N = 128$, the author got an error of about 0.043, using either node set. The problem is the very sharp "kink" at $x = 0$, as shown in the plots below for the Lobatto nodes (Figs 4.18 and 4.19).

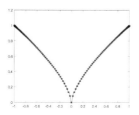

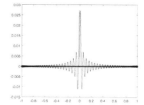

Figure 4.30   **Figure 4.31**

15. For $V = 50$, define the function

$$f(x) = 1 + Ae^{ax} + Be^{bx},$$

where a is the larger root of the polynomial $p(x) = -r^2 + Vr = 1$ and b is the smaller root; $\delta = e^{b-a} - e^{a-b}$, and

$$A = -e^b/\delta, \qquad B = e^a/\delta.$$

Plot this function to see what it looks like, then construct a spectral interpolant (using either set of nodes). How large do you need to take N to get an error that is less than 10^{-5}? What about 10^{-12}?

16. For $y = \sqrt{x}$ and using $N = 2$, employ the change of variable from Problem 7 in §4.11.3 to construct interpolants using the Lobatto and Chebyshev nodes. Don't write a script to do this, but feel free to use reasonable computational tools to simplify the work.

17. Modify your script from Problem 13 to work on any interval $[a, b]$, using the change of variables from Problem 7 in §4.11.3. Make sure your final result will work on the interval $[a, b]$.

18. Prove Theorem 4.13.

CHAPTER 5

NUMERICAL INTEGRATION

5.1 A REVIEW OF THE DEFINITE INTEGRAL

Exercises:

1. Basic properties of the definite integral show that it is a *linear operator*, i.e., it distributes across sums and multiplication by constants:

$$I(\alpha f + \beta g) = \alpha I(f) + \beta I(g).$$

Prove that if

$$I_n(f) = \sum_{i=0}^{n} w_i f(x_i)$$

then I_n is also linear:

$$I_n(\alpha f + \beta g) = \alpha I_n(f) + \beta I_n(g).$$

Solutions Manual to Accompany An Introduction to Numerical Methods and Analysis, Third Edition.
James F. Epperson.
© 2021 John Wiley & Sons, Inc. Published 2021 by John Wiley & Sons, Inc.

Solution: This is a straight-forward computation:

$$
\begin{aligned}
I_n(\alpha f + \beta g) &= \sum_{i=0}^{n} w_i(\alpha f(x_i) + \beta g(x_i)) \\
&= \sum_{i=0}^{n} w_i \alpha f(x_i) + \sum_{i=0}^{n} \beta g(x_i) \\
&= \alpha \sum_{i=0}^{n} w_i f(x_i) + \beta \sum_{i=0}^{n} g(x_i) \\
&= \alpha I_n(f) + \beta I_n(g).
\end{aligned}
$$

2. Assume that the quadrature rule I_n integrates all polynomials of degree less than or equal to N exactly:

$$
I_n(p) = I(p)
$$

for all $p \in \mathcal{P}_N$. Use this to prove that, for any integrand f, the error $I - I_n$ is equal to the error in integrating the Taylor remainder:

$$
I(f) - I_n(f) = I(R_N) - I_n(R_N),
$$

where $f(x) = p_N(x) + R_N(x)$. Does it really matter that we are using the Taylor polynomial and remainder? In other words, will this result hold for any polynomial approximation and its associated error?

Solution: The previous problem established the linearity of I_n, which is the important issue. We have

$$
\begin{aligned}
I(f) - I_n(f) &= I(p_N + R_N) - I_n(I_N + R_N) \\
&= I(p_N) + I(R_N) - I_n(p_N) - I_n(R_N) \\
&= I(R_N) - I_n(R_N).
\end{aligned}
$$

3. Use Problem 1 to prove the following: If a quadrature rule I_n is exact for all powers x^k for $k \leq d$, then it is exact for all polynomials of degree less than or equal to d.

Solution: Let p be an arbitrary polynomial of degree less than or equal to d, and let $\xi_k(x) = x^k$. Then we can write

$$
p(x) = \sum_{k=0}^{d} a_k \xi_k(x).
$$

The linearity that was proved in Problem 1 allows us to write

$$
I_n(p) = I_n \left(\sum_{k=0}^{d} a_k \xi_k \right) = \sum_{k=0}^{d} a_k I_n(\xi_k).
$$

Therefore, $I(p) = I_n(p)$ if and only if $I(xi_k) = I_n(\xi_k)$ for all k.

◁ • • • ▷

5.2   IMPROVING THE TRAPEZOID RULE

Exercises:

1. Apply the trapezoid rule and corrected trapezoid rule, with $h = \frac{1}{4}$, to approximate the integral

$$I = \int_0^1 x(1 - x^2)dx = \frac{1}{4}.$$

Solution: We have

$$T_4(f) = (0.25/2)(0 + 2(0.234375) + 2(0.375) + 2(0.328125) + 0) = 0.234375.$$

Also,

$$f'(x) = 1 - 3x^2,$$

so $f'(0) = 1$ and $f'(1) = -2$. Therefore,

$$T_4^C(f) = T_4(f) - (0.25^2/12)(-2 - 1) = 0.25,$$

which is exact.

2. Apply the trapezoid rule and corrected trapezoid rule, with $h = \frac{1}{4}$, to approximate the integral

$$I = \int_0^1 \frac{1}{\sqrt{1 + x^4}}dx = 0.92703733865069.$$

Solution: The trapezoid rule gives us

$$
\begin{aligned}
T_4(f) &= (0.25/2)(1 + (2/\sqrt{1.00390625}) + (2/\sqrt{1.0625}) + (2/\sqrt{1.31640625}) + (1/\sqrt{2})) \\
&= 0.9233310015.
\end{aligned}
$$

The corrected trapezoid value is then computed from this very simply:

$$
\begin{aligned}
T_4^C(f) &= 0.9233310015 - (0.0625/12)(f'(1) - f'(0)) \\
&= 0.9233310015 - (5.208333333 \times 10^{-3})(-2/2^{3/2}) \\
&= 0.9270138493,
\end{aligned}
$$

which is much more accurate than the original T_4 value.

3. Apply the trapezoid rule and corrected trapezoid rule, with $h = \frac{1}{4}$, to approximate the integral

$$I = \int_0^1 \ln(1 + x)dx = 2\ln 2 - 1.$$

Solution: We have

$$T_4(f) = (0.25/2)(0 + 2(0.22314355) + 2(0.40546511) + 2(0.55961579) + 0.69314718)$$
$$= 0.38369951.$$

We have

$$f'(x) = 1/(1 + x),$$

so $f'(0) = 1$ and $f'(1) = 1/2$. Therefore,

$$T_4^C(f) = T_4(f) - (0.25^2/12)(0.5-1) = 0.3836995094 + (0.0625/24) = 0.3863036761.$$

The exact value is 0.386294361.

4. Apply the trapezoid rule and corrected trapezoid rule, with $h = \frac{1}{4}$, to approximate the integral

$$I = \int_0^1 \frac{1}{1+x^3}\,dx = \frac{1}{3}\ln 2 + \frac{1}{9}\sqrt{3}\pi.$$

Solution: We have

$$
\begin{aligned}
T_4(f) &= (.25/2)(1 + 2(0.9846153846) + 2(0.8888888889) + 2(0.7032967033) + 0.5) \\
&= 0.8317002443.
\end{aligned}
$$

Then we have

$$f'(x) = -3x^2/(1+x^3)^2,$$

so $f'(0) = 0$ and $f'(1) = -3/4$. Therefore,

$$T_4^C(f) = T_4(f) - (0.25^2/12)(-0.75 - 0) = 0.8356064943.$$

The exact value is 0.8356488485.

5. Apply the trapezoid rule and corrected trapezoid rule, with $h = \frac{1}{4}$, to approximate the integral

$$I = \int_1^2 e^{-x^2}\,dx = 0.1352572580.$$

Solution: We have

$$
\begin{aligned}
T_4(f) &= (0.25/2)(0.3678794412 + 2(0.2096113872) + 2(0.1053992246) \\
&\quad + 2(0.4677062238 \times 10^{-1}) + 0.1831563889 \times 10^{-1}) \\
&= 0.1387196936.
\end{aligned}
$$

We have

$$f'(x) = -2xe^{-x^2},$$

so $f'(1) = -1/e$ and $f'(2) = -1/e^4$. Therefore,

$$T_4^C(f) = T_4(f) - (0.25^2/12)(-e^{-4} + e^{-1}) = 0.1352691919.$$

The exact value is 0.1352572580.

6. For each integral below, write a program to do the corrected trapezoid rule using the sequence of mesh sizes $h = \frac{1}{2}(b-a), \frac{1}{4}(b-a), \frac{1}{8}(b-a), \ldots, \frac{1}{2048}(b-a)$, where $b - a$ is the length of the given interval. Verify that the expected rate of decrease of the error is observed.

   (a) $f(x) = x^2 e^{-x}$, $[0, 2]$, $I(f) = 2 - 10e^{-2} = 0.646647168$;
   (b) $f(x) = 1/(1+x^2)$, $[-5, 5]$, $I(f) = 2\arctan(5)$;
   (c) $f(x) = \ln x$, $[1, 3]$, $I(f) = 3\ln 3 - 2 = 1.295836867$;
   (d) $f(x) = e^{-x}\sin(4x)$, $[0, \pi]$, $I(f) = \frac{4}{17}(1 - e^{-\pi}) = 0.2251261368$;
   (e) $f(x) = \sqrt{1-x^2}$, $[-1, 1]$, $I(f) = \pi/2$.

Solution: For the single case of (c), my program produced the output in Table 5.1.

Table 5.1   Data for Problem 6.

n	T_n	T_n^C	$T_n - T_n^C$	$I - T_n^C$	Ratio
2	1.24245332	1.29800888	0.555556×10^{-1}	-0.217201×10^{-2}	0.0000
4	1.28210458	1.29599347	0.138889×10^{-1}	-0.156605×10^{-3}	13.8694
8	1.29237491	1.29584713	0.347222×10^{-2}	-0.102642×10^{-4}	15.2574
16	1.29496946	1.29583752	0.868056×10^{-3}	-0.650071×10^{-6}	15.7894
32	1.29561989	1.29583691	0.217014×10^{-3}	-0.407687×10^{-7}	15.9453
64	1.29578262	1.29583687	0.542535×10^{-4}	-0.255025×10^{-8}	15.9862
128	1.29582330	1.29583687	0.135634×10^{-4}	-0.159425×10^{-9}	15.9965
256	1.29583348	1.29583687	0.339084×10^{-5}	$-0.996447 \times 10^{-11}$	15.9994
512	1.29583602	1.29583687	0.847711×10^{-6}	$-0.623501 \times 10^{-12}$	15.9815
1024	1.29583665	1.29583687	0.211928×10^{-6}	$-0.384137 \times 10^{-13}$	16.2312

7. Apply the trapezoid rule and corrected trapezoid rule to the approximation of

$$I = \int_0^1 x^2 e^{-2x} dx = 0.0808308960... \quad .$$

Compare your results in the light of the expected error theory for both methods, and comment on what occurs. How does the error behave in each case, as a function of h? How should it have behaved?

Solution: When doing the computation we find that $f'(0) - f'(1) = 0$, thus the "correction term" will be zero, thus the original trapezoid rule will be just as accurate as the corrected rule.

8. Repeat the above for

$$I = \int_0^\pi \sin^2 x dx = \frac{1}{2}\pi.$$

Solution: The same thing happens.

9. The length of a curve $y = g(x)$, for x between a and b, is given by the integral

$$L(g) = \int_a^b \sqrt{1 + [g'(x)]^2} dx.$$

Use the corrected trapezoid rule to find the length of one "arch" of the sine curve.

Solution: This amounts to using the corrected trapezoid rule to evaluate the integral

$$L = \int_0^\pi \sqrt{1 + \cos^2(x)} dx.$$

Using 16 trapezoids we get $L \approx 3.820197788$, which is accurate to all digits displayed.

10. Use the corrected trapezoid rule to find the length of the exponential function from $x = -1$ to $x = 1$. How small does h have to be for the computation to converge to within 10^{-6}?

11. Repeat the above for the tangent function, from $x = -\pi/4$ to $x = \pi/4$.

12. Define the function

$$F(t) = \int_a^t f(x)dx$$

and note that

$$F(b) = I(f) = \int_a^b f(x)dx.$$

Use Taylor expansions of F and f about $x = a$ to show that

$$I(f) - \frac{1}{2}(b-a)(f(b) + f(a)) - \frac{1}{12}(b-a)^2(f'(b) - f'(a)) = \mathcal{O}((b-a)^5).$$

Use this to show that the corrected trapezoid rule is $\mathcal{O}(h^4)$ when applied over a uniform grid of length h.

13. Construct a version of the "quasi-corrected" trapezoid rule that uses the derivative approximations

$$f'(a) \approx \frac{f(a+h) - f(a)}{h}, \quad f'(b) \approx \frac{f(b) - f(b-h)}{h}.$$

Explain why we should expect this to be less accurate than the rule using the approximations (5.2) and (5.3), and demonstrate that this is the case on the integral

$$I = \int_0^1 x(1 - x^2)dx = \frac{1}{4}.$$

Solution: The approximations in (5.2) and (5.3) are $\mathcal{O}(h^2)$ accurate, whereas the approximations suggested in this exercise are only $\mathcal{O}(h)$. We thus expect to get less accuracy when using these approximations in the corrected trapezoid rule, and this is borne out by actual experiment. For the example in this exercise, the corrected trapezoid rule using the $\mathcal{O}(h^2)$ formulas is *exact*, but the approximation using the less accurate formulas suggested in this exercise still shows some error for $h = (1/2)^5$.

5.3 SIMPSON'S RULE AND DEGREE OF PRECISION

Exercises:

1. Apply Simpson's rule with $h = \frac{1}{4}$, to approximate the integral

$$I = \int_0^1 x(1 - x^2)dx = \frac{1}{4}.$$

Solution: We have

$$S_4(f) = (0.25/3)(0+4(0.2343750000)+2(0.3750000000)+4(0.3281250000)+0) = 0.25,$$

which is exact.

2. Apply Simpson's rule with $h = \frac{1}{4}$, to approximate the integral

$$I = \int_0^1 \frac{1}{\sqrt{1+x^4}} dx = 0.92703733865069.$$

Solution: We have

$$S_4(f) = (0.25/3)(1 + 4(0.9980525784) + 2(0.9701425000) + 4(0.8715755371)$$
$$+ 0.7071067810)$$
$$= 0.9271586870.$$

3. Apply Simpson's rule with $h = \frac{1}{4}$, to approximate the integral

$$I = \int_0^1 \ln(1+x)dx = 2\ln 2 - 1.$$

How small does the error theory say that h has to be to get that the error is less than 10^{-3}? 10^{-6}? How small does h have to be for the trapezoid rule to achieve this accuracy?

Solution: For $h = 1/4$, we get $S_4(f) = 0.3862595628$. To get the error less than 10^{-3} we need

$$\frac{h^4}{180} \times \frac{6}{(1+0)^4} \le 10^{-3},$$

which implies that $h \le 0.4161791450$. To get 10^{-6} accuracy, we need $h \le 0.7400828045 \times 10^{-1}$. Using the trapezoid rule, we need to have

$$\frac{h^2}{12} \times \frac{1}{(1+0)^2} \le 10^{-3},$$

so that we need $h \le 0.1095445115$ and $h \le 0.3464101615 \times 10^{-2}$.

4. Apply Simpson's rule with $h = \frac{1}{4}$, to approximate the integral

$$I = \int_0^1 \frac{1}{1+x^3} dx = \frac{1}{3}\ln 2 + \frac{1}{9}\sqrt{3}\pi.$$

Solution: We have

$$S_4(f) = (0.25/3)(1 + 4(0.9846153846) + 2(0.8888888889) + 4(0.7032967033) + 0.5)$$
$$= 0.8357855108.$$

The exact value is 0.8356488485.

5. Apply Simpson's rule with $h = \frac{1}{4}$, to approximate the integral

$$I = \int_1^2 e^{-x^2} dx = 0.1352572580.$$

Solution: We have

$$S_4(f) = (0.25/3)(0.3678794412 + 4(0.2096113872) + 2(0.1053992246)$$
$$+ 4(0.4677062238 \times 10^{-1}) + 0.1831563889 \times 10^{-1})$$
$$= 0.1352101306.$$

6. For each function below, write a program to do Simpson's rule using the sequence of mesh sizes $h = \frac{1}{2}(b-a), \frac{1}{4}(b-a), \frac{1}{8}(b-a), \ldots, \frac{1}{2048}(b-a)$, where $b-a$ is the length of the given interval. Verify that the expected rate of decrease of the error is observed. Comment on any anomolies that are observed.

   (a) $f(x) = \ln x$, $[1,3]$, $I(f) = 3\ln 3 - 2 = 1.295836867$;
   (b) $f(x) = x^2 e^{-x}$, $[0,2]$, $I(f) = 2 - 10e^{-2} = 0.646647168$;
   (c) $f(x) = 1/(1+x^2)$, $[-5,5]$, $I(f) = 2\arctan(5)$;
   (d) $f(x) = \sqrt{1-x^2}$, $[-1,1]$, $I(f) = \pi/2$;
   (e) $f(x) = e^{-x}\sin(4x)$, $[0,\pi]$, $I(f) = \frac{4}{17}(1-e^{-\pi}) = 0.2251261368.$

   Solution: My program produced the output (for (a)–(d), only) in Table 5.2. Note that the accuracy is as it should be for all but the single case of (d), for which the error theory does not apply because of the singularity in the derivative at each endpoint.

Table 5.2   Results for Exercise 5.3.6.

n	(c)	(b)	(a)	(d)
2	6.79487179487180	0.67095296587741	1.29040033696930	1.33333333333333
4	2.65030946065429	0.64863368605631	1.29532166828621	1.48803387171258
8	2.61738281929855	0.64678014940774	1.29579834986087	1.54179757747348
16	2.73331903435335	0.64665562384530	1.29583431135754	1.56059458488771
32	2.74671071699175	0.64664769843876	1.29583670367239	1.56719883449230
64	2.74680148839078	0.64664720084510	1.29583685581508	1.56952610894680
128	2.74680153128328	0.64664716971014	1.29583686536681	1.57034753804027
256	2.74680153372702	0.64664716776365	1.29583686596447	1.57063770946780
512	2.74680153387984	0.64664716764198	1.29583686600184	1.57074025657205
1024	2.74680153388940	0.64664716763438	1.29583686600417	1.57077650465356
2048	2.74680153388999	0.64664716763390	1.29583686600432	1.57078931890607

7. For each integral in the previous problem, how small does h have to be to get accuracy, according to the error theory, of at least 10^{-3}? 10^{-6}? Compare to the value of h required by the trapezoid rule for this accuracy. (Feel free to use a computer algebra system to help you with the computation of the derivatives.)

Solution:

(a) For $f(x) = \ln x$ we have

$$|I(f) - S_n(f)| \leq \frac{2h^4}{180} \max_{x \in [1,3]} \left| \frac{-6}{x^4} \right| \leq \frac{h^4}{90} \times 6 = \frac{h^4}{15}.$$

Therefore we require $h \leq 0.3499635512$ for 10^{-3} accuracy and $h \leq 0.6223329773 \times 10^{-1}$ for 10^{-6}.

(b) For $f(x) = x^2 e^{-x}$ we have

$$|I(f) - S_n(f)| \leq \frac{2h^4}{180} \max_{x \in [0,2]} \left| (x^2 - 8x + 12)e^{-x} \right| \leq \frac{h^4}{90} \times 12 = \frac{2h^4}{15}.$$

Therefore we require $h \leq 0.2942830956$ for 10^{-3} accuracy and $h \leq 0.5233175697 \times 10^{-1}$ for 10^{-6}.

(c) For $f(x) = 1/(1 + x^2)$, we have

$$|I(f) - S_n(f)| \leq \frac{10h^4}{180} \max_{x \in [-5,5]} \left| \frac{24(5x^4 - 10x^2 + 1)}{(1 + x^2)^5} \right| \leq \frac{4h^4}{3} \times 24 = 32h^4.$$

Therefore we require $h \leq 0.7476743906 \times 10^{-1}$ for 10^{-3} accuracy and $h \leq 0.1329573974 \times 10^{-1}$ for 10^{-6}.

(d) For $f(x) = \sqrt{1 - x^2}$ we have that the fourth derivative is not bounded on $[-1, 1]$ so the error theory does not apply.

(e) For $f(x) = e^{-x} \sin(4x)$ we have

$$|I(f) - S_n(f)| \leq \frac{\pi h^4}{180} \max_{x \in [0,\pi]} \left| (161 \sin 4x + 240 \cos 4x)e^{-x} \right|$$

$$\leq \frac{\pi h^4}{180} \times 260 = \frac{13\pi h^4}{9}.$$

Therefore we require $h \leq 0.1218392792$ for 10^{-3} accuracy and $h \leq 0.2166642816 \times 10^{-1}$ for 10^{-6}.

8. Since the area of the unit circle is $A = \pi$, it follows that

$$\frac{\pi}{2} = \int_{-1}^{1} \sqrt{1 - x^2} dx.$$

Therefore, we can approximate π by approximating this integral. Use Simpson's rule to compute approximate values of π in this way and comment on your results.

Solution: Using the sequence of h values $h = 1, 1/2, 1/4, 1/8, \ldots$, we get the following table of approximate values of π.

Estimate of π
2.666666667
2.976067744
3.083595156
3.121189170
3.134397670
3.139052218
3.140695076
3.141275419
3.141480514
3.141553009
3.141578638
3.141587698

Clearly the values are converging to π, but not as rapidly as we would expect from Simpson's rule. The reason is that the integrand is not smooth at either endpoint, thus the error estimate for Simpson's rule does not apply.

9. If we wanted to use Simpson's rule to approximate the natural logarithm function on the interval $[\frac{1}{2}, 1]$ by approximating

$$\ln x = \int_1^x \frac{1}{t} dt$$

how many points would be needed to obtain an error of less than 10^{-6}? How many points for an error of less than 10^{-16}? What are the corresponding values for the trapezoid rule?

Solution: The error using Simpson's rule is bounded above according to

$$E \leq \frac{(1/2)h^4}{180} \max_{x \in (1/2),1} \frac{24}{x^5} = \frac{32h^4}{15}.$$

Therefore, to get 10^{-6} accuracy requires imposing

$$32h^4/15 \leq 10^{-6},$$

which implies $h \leq 0.026166$, therefore the number of required points is $n = 20$. To get 10^{-16} accuracy requires $h \leq 0.8274377295 \times 10^{-4}$ and $n = 6,043$. For the trapezoid rule we get $h \leq 0.0049$ and $n \geq 103$, and $h \leq 0.4899 \times 10^{-7}$ and $n \geq 10,206,208$. Clearly Simpson's rule is more efficient.

10. Use Simpson's rule to produce a graph of $E(x)$, defined to be the *length* of the exponential curve from 0 to x, for $0 \leq x \leq 3$. See Problem 14 of §2.5.

Solution: If we use a fixed number of points, say $n = 128$, for any value of x, then we get the plot shown in Figure 5.1.

11. Let $f(x) = |x|$; use the trapezoid rule (with $n = 1$), corrected trapezoid rule (with $n = 1$), and Simpson's rule (with $n = 2$), to compute

$$I(f) = \int_{-1}^1 f(x) dx$$

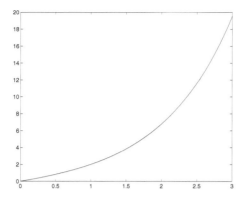

Figure 5.1   Solution for Exercise 5.3.10.

and compare your results to the exact value. Explain what happens in the light of our error estimates for the trapezoid and Simpson's rules.

Solution: Because the function is piecewise linear, and one grid point matches the "kink" in the graph, the trapezoid rule will produce the exact value of the integral, but neither Simpson's rule nor the corrected trapezoid rule will be exact.

12. Write out the expression for Simpson's rule when c is *not* the midpoint of the interval $[a, b]$. To simplify matters, take $c - a = h$, $b - c = \theta h$.

Solution: We get

$$S_2(f) = (h/6)(Af(a) + Cf(c) + Bf(b)),$$

where

$$A = 2 + \theta - \theta^2, \quad C = (3\theta^2 + 1 + \theta^3 + 3\theta)/\theta, \quad B = (2\theta^2 + \theta - 1)/\theta.$$

13. What is the degree of precision of the corrected trapezoid rule, T_1^C? What about the n subinterval version, T_n^C?

Solution: Let $\xi_k(x) = x^k$, and consider the computation

$$
\begin{aligned}
I_k &= T_1^C(\xi_k) = \frac{1}{2}(b - a)\left(a^k + b^k\right) - \frac{b - a}{12}\left(kb^{k-1} - ka^{k-1}\right) \\
&= \frac{1}{2}(b - a)(b^k + a^k) - \frac{1}{12}(b - a)^2(kb^{k-1} - ka^{k-1}).
\end{aligned}
$$

When simplified, and compared to the exact values of

$$I = \int_a^b \xi_k(x)dx$$

we get that $I_k = I$ for $k = 0, 1, 2, 3$, but not for $k = 4$, thus showing that the degree of precision is $p = 3$. The same result holds for the n subinterval version.

14. Prove that if we want to show that the quadrature rule $I_n(f)$ has degree of precision p, it suffices to show that it will exactly integrate x^k, $0 \leq k \leq p$ over the integral $(0, 1)$.

    Solution: Let $q(x)$ be an arbitrary polynomial of degree p. Then

    $$q(x) = \sum_{k=0}^{p} a_k x^k,$$

    so

    $$I(q) = \sum_{k=0}^{p} \int_a^b x^k \, dx.$$

    Therefore, we need only concern ourselves with exactly integrating simple powers instead of general polynomials. A change of variable can now be employed to transform the interval of interest to $[0, 1]$ from the more general $[a, b]$.

15. Construct the analogue of Simpson's rule based on exactly integrating a cubic interpolate at equally spaced points on the interval $[a, b]$.

    Solution: This yields what is sometimes known as Simpson's three-eighths rule:

    $$I_3(f) = \frac{3h}{8} \left(f(a) + 3f(a + h) + 3f(a + 2h) + f(a + 3h) \right).$$

16. Show that $I(q_3) = S_2(f)$.

    Solution: This is actually a straight-forward (if lengthy) computation with $q_3(x)$ as defined here:

    $$q_3(x) = f(a) + \left(\frac{f(c) - f(a)}{h} \right)(x - a)$$
    $$+ \left(\frac{f'(c) - \frac{f(c) - f(a)}{h}}{h} \right)(x - a)(x - c) + A(x - a)(x - c)^2,$$

    where

    $$A = h^{-2} \left(\frac{f(b) - f(c)}{h} - 2f'(c) + \frac{f(c) - f(a)}{h} \right)$$

    and

    $$h = c - a = b - c.$$

17. Consider the quadrature rule defined by exactly integrating a cubic Hermite interpolate:

    $$I_1(f) = I(H_2).$$

    Write down the quadrature formula for both the basic and composite settings, and state and prove an error estimate, using the error results for Hermite interpolation from the previous chapter.

    Solution: The Hermite cubic interpolant can be written as

    $$H_2(x) = h_a(x)f(a) + h_b(x)f(b) + \tilde{h}_a(x)f'(a) + \tilde{h}_b(x)f'(b),$$

where the h and $\tilde{h}$ functions are given in §4.6 as follows:

$$h_a(x) = \left(1 - 2\left(\frac{x-a}{a-b}\right)\right)\left(\frac{x-b}{a-b}\right)^2,$$

$$h_b(x) = \left(1 - 2\left(\frac{x-b}{b-a}\right)\right)\left(\frac{x-a}{b-a}\right)^2,$$

$$\tilde{h}_a(x) = (x-a)\left(\frac{x-b}{a-b}\right)^2,$$

$$\tilde{h}_b(x) = (x-b)\left(\frac{x-a}{b-a}\right)^2.$$

Then the integration rule is

$$I_1(f) = Af(a) + Bf(b) + \alpha f'(a) + \beta f'(b),$$

where

$$A = \int_a^b h_a(x)dx = \frac{1}{2}(b-a),$$

$$B = \int_a^b h_b(x)dx = \frac{1}{2}(b-a),$$

$$\alpha = \int_a^b \tilde{h}_a(x)dx = \frac{1}{12}(b-a)^2,$$

$$\beta = \int_a^b \tilde{h}_a(x)dx = -\frac{1}{12}(b-a)^2.$$

The n interval rule is easily derived from the single interval rule. The error estimate is found as follows:

$$\begin{aligned} I(f) - I_1(f) &= \int_a^b (f(x) - H_2(x))\,dx \\ &= \frac{1}{24}\int_a^b (x-a)^2(x-b)^2 f^{(4)}(\xi_x)dx \\ &= \left(\frac{1}{24}\int_a^b (x-a)^2(x-b)^2 dx\right) f^{(4)}(\xi) \\ &= \frac{(b-a)^5}{720}f^{(4)}(\xi). \end{aligned}$$

18. Consider a quadrature rule in the form

$$I_n(f) = \sum_{k=1}^n a_k f(x_k),$$

where the coefficients $a_k > 0$ and the grid points x_k are all known. Assume that I_n integrates the trivial function $w(x) = 1$ exactly:

$$I_n(w) = \sum_{k=1}^n a_k = I(w) = \int_a^b dx = b - a,$$

and that this holds for all intervals (a, b). Consider now the effects of rounding error on integrating an arbitrary function f. Let $\hat{f}(x) = f(x) + \epsilon(x)$ be f polluted by rounding error, with $|\epsilon(x)| \leq C\mathbf{u}$ for some constant $C > 0$, for all $x \in [a, b]$. Show that

$$|I_n(f) - I_n(\hat{f})| \leq C\mathbf{u}(b - a).$$

Comment on this in comparison to the corresponding result for numerical differentiation, as given in §2.2.

Solution: The point is that numerical integration is much less affected by rounding error than is numerical differentiation.

19. The *normal probability distribution* is defined as

$$p(x) = \frac{1}{\sigma\sqrt{2\pi}}e^{-(x-\mu)^2/2\sigma^2},$$

where μ is the mean, or average, and σ is the variance. This is the famous bell-shaped curve that one hears so much about; the mean gives the center of the bell and the variance gives its width. If x is distributed in this fashion, then the probability that $a \leq x \leq b$ is given by the integral

$$P(a \leq x \leq b) = \int_a^b p(x)dx.$$

(a) Use the change of variable $z = (x - \mu)/\sigma$ to show that

$$P(-m\sigma \leq x \leq m\sigma) = \frac{1}{\sqrt{2\pi}}\int_{-m}^m e^{-z^2/2}dz.$$

(b) Compute values of $P(-m\sigma \leq x \leq m\sigma)$ for $m = 1, 2, 3$, using Simpson's rule.

Solution: For $m = 1$, $P(-m\sigma \leq x \leq m\sigma) \approx 0.6826908120$; for $m = 2$, $P(-m\sigma \leq x \leq m\sigma) \approx 0.9544947255$; and $m = 3$, $P(-m\sigma \leq x \leq m\sigma) \approx 0.9972830690$.

20. Use Simpson's rule to solve Problem 9 from §5.2.

21. Use Simpson's rule to solve Problem 10 from §5.2.

◁ • • • ▷

5.4 THE MIDPOINT RULE

Exercises:

1. Apply the midpoint rule with $h = \frac{1}{4}$, to approximate the integral

$$I = \int_0^1 x(1 - x^2)dx = \frac{1}{4}.$$

How small does h have to be to get that the error is less than 10^{-3}? 10^{-6}?

Solution: We have, for $f(x) = x(1 - x^2)$,

$$M_4(f) = (1/4)(f(1/8) + f(3/8) + f(5/8) + f(7/8)) = 0.2578125000.$$

The error estimate is

$$|I(f) - M_n(f)| \leq \frac{h^2}{24} \max_{x \in [0,1]} |-6x| = \frac{h^2}{4},$$

so we require $h \leq 0.6324555320 \times 10^{-1}$ for 10^{-3} accuracy, and $h \leq 0.002$ for 10^{-6}.

2. Apply the midpoint rule with $h = \frac{1}{4}$, to approximate the integral

$$I = \int_0^1 \frac{1}{\sqrt{1 + x^4}} dx = 0.92703733865069.$$

How small does h have to be to get that the error is less than 10^{-3}? 10^{-6}?

Solution:
$$M_4(f) = 0.9288993588,$$

and $h \leq 0.1312709324$ for 10^{-3} accuracy, $h \leq 0.4151151371e - 2$ for 10^{-6}.

3. Apply the midpoint rule with $h = \frac{1}{4}$, to approximate the integral

$$I = \int_0^1 \ln(1 + x) dx = 2\ln 2 - 1.$$

How small does h have to be to get that the error is less than 10^{-3}? 10^{-6}?

Solution:
$$M_4(f) = 0.3875883105,$$

and $h \leq 0.1549193338$ for 10^{-3} accuracy, $h \leq 0.4898979486 \times 10^{-2}$ for 10^{-6}.

4. Apply the midpoint rule with $h = \frac{1}{4}$, to approximate the integral

$$I = \int_0^1 \frac{1}{1 + x^3} dx = \frac{1}{3}\ln 2 + \frac{1}{9}\sqrt{3}\pi.$$

How small does h have to be to get that the error is less than 10^{-3}? 10^{-6}?

Solution:
$$M_4(f) = 0.8376389970,$$

and $h \leq 0.1175282713$ for 10^{-3} accuracy, $h \leq 0.3716570267 \times 10^{-2}$ for 10^{-6}.

5. Apply the midpoint rule with $h = \frac{1}{4}$, to approximate the integral

$$I = \int_1^2 e^{-x^2} dx = 0.1352572580.$$

How small does h have to be to get that the error is less than 10^{-3}? 10^{-6}?

Solution:

$$M_4(f) = 0.1335215673,$$

and $h \leq 0.1095445115$ for 10^{-3} accuracy, $h \leq 0.3464101615 \times 10^{-2}$ for 10^{-6}.

6. Show that the midpoint rule can be derived by integrating, exactly, a polynomial interpolate of degree *zero*.

   Solution: Take $p_0(x) = f(c)$ where $c = (a + b)/2$ is the midpoint of the interval. Then

   $$I(p_0) = \int_a^b p_0(x)dx = f(c)(b - a) = M_1(f).$$

7. Apply the midpoint rule to each of the following functions, integrated over the indicated interval. Use a sequence of grids $h = (b - a), (b - a)/2, (b - a)/4, \ldots$ and confirm that the approximations are converging at the correct rate. Comment on any anomolies that you observe.

   (a) $f(x) = \ln x$, $[1, 3]$, $I(f) = 3\ln 3 - 2 = 1.295836867$;

   (b) $f(x) = x^2 e^{-x}$, $[0, 2]$, $I(f) = 2 - 10e^{-2} = 0.646647168$;

   (c) $f(x) = \sqrt{1 - x^2}$, $[-1, 1]$, $I(f) = \pi/2$;

   (d) $f(x) = 1/(1 + x^2)$, $[-5, 5]$, $I(f) = 2\arctan(5)$;

   (e) $f(x) = e^{-x}\sin(4x)$, $[0, \pi]$, $I(f) = \frac{4}{17}(1 - e^{-\pi}) = 0.2251261368$.

   Solution: The output is in Table 5.3 ((a)–(d), only). The approximation performed as expected, with two notable exceptions: For $f(x) = \sqrt{1 - x^2}$, we did not achieve the expected rate of decrease for the error, because the second derivative of f is unbounded on $[-1, 1]$. For $f(x) = x^2 e^{-x}$, we achieved *higher* accuracy than expected. This occurs because $f'(0) = f'(2)$ and there is an effect similar to the corrected trapezoid rule.

Table 5.3   Results for Exercise 5.4.7.

n	(d)	(b)	(a)	(c)
2	1.37931034482759	0.65367552526213	1.32175583998232	1.73205080756888
4	2.28316971966400	0.64711384088491	1.30264523357222	1.62968366431800
8	2.70773386697212	0.64667678393337	1.29756401303372	1.59196461030595
16	2.74701026646563	0.64664902575751	1.29627032524886	1.57834346564915
32	2.74692176145878	0.64664728387795	1.29594533728055	1.57347590396326
64	2.74683162099652	0.64664717490087	1.29586399050904	1.57174570134112
128	2.74680905737720	0.64664716808809	1.29584364754886	1.57113233600677
256	2.74680341486880	0.64664716766226	1.29583856141662	1.57091518675892
512	2.74680200414141	0.64664716763565	1.29583728985904	1.57083836124104
1,024	2.74680165145330	0.64664716763398	1.29583697196811	1.57081119017919
2,048	2.74680156328087	0.64664716763388	1.29583689249528	1.57080158214194

8. For each integral in the previous problem, how small does h have to be to get accuracy, according to the error theory, of at least 10^{-3}? 10^{-6}?

Solution: For (b), we have

$$|I(f) - M_n(f)| \le \frac{10h^2}{24} \max_{x \in [-5,5]} \left| \frac{6x^2 - 2}{(1 + x^2)^3} \right| = \frac{5h^2}{6},$$

therefore we need $h \le 0.3464101615 \times 10^{-1}$ to get 10^{-3} accuracy, and $h \le 0.1095445115 \times 10^{-2}$ for 10^{-6} accuracy.

9. State and prove a formal theorem concerning the error estimate for the midpoint rule over n subintervals. You may want to state and prove a formal theorem for the single subinterval rule first, and then use this in the more general theorem.

   Solution: The theorem would go something like this:

   Theorem: Let $f \in C^2([a, b])$; then there exists a point $\xi \in [a, b]$ such that

   $$I(f) - M_n(f) = -\frac{b-a}{24} h^2 f''(\xi).$$

   The proof is almost identical to the developments used for Simpson's rule and the trapezoid rule.

10. Let T_1 be the trapezoid rule using a single subinterval, M_1 be the midpoint rule using a single subinterval, and S_2 be Simpson's rule using a single quadratic interpolant (hence, a single *pair* of subintervals). Show that, for any continuous function f, and any interval $[a, b]$,

    $$S_2(f) = \frac{1}{3}T_1(f) + \frac{2}{3}M_1(f).$$

    Solution: Let c denote the midpoint of the interval $[a, b]$, so that we have

    $$T_1(f) = \frac{1}{2}(b - a)(f(b) + f(a)),$$

    $$S_2(f) = \frac{(b-a)/2}{3}(f(a) + 4f(c) + f(b)),$$

    $$M_1(f) = (b - a)f(c).$$

    Then a direct computation shows

    $$\begin{aligned}
    \frac{1}{3}T_1(f) + \frac{2}{3}M_1(f) &= \frac{1}{6}(b-a)(f(b) + f(a)) + \frac{2}{3}(b-a)f(c) \\
    &= \frac{(b-a)/2}{3}(f(a) + 4f(c) + f(b)) = S_2(f).
    \end{aligned}$$

◁ • • • ▷

5.5 APPLICATION: STIRLING'S FORMULA

Exercises:

1. Use Stirling's formula to show that

$$\lim_{n \to \infty} \left(\frac{x^n}{n!} \right) = 0$$

for all x.

Solution: We have

$$\left| \frac{x^n}{n!} \right| \leq \frac{|x^n|}{C_n \sqrt{n}(n/e)^n} = \left(\frac{1}{C_n \sqrt{n}} \right) \left(\frac{|x|e}{n} \right)^n.$$

For n sufficiently large, the second factor will be less than one, which completes the proof.

2. For $x = 10$ and $\epsilon = 10^{-3}$, how large does n have to be for

$$\left| \frac{x^n}{n!} \right| \leq \epsilon$$

to hold? Repeat for $\epsilon = 10^{-6}$. Use Stirling's formula here, don't just plug numbers into a calculator.

Solution: We have, for all $n \geq 1$,

$$\frac{10^n}{n!} = \frac{10^n e^n}{C_n \sqrt{n} n^n} \leq \frac{1}{2} \left(\frac{10e}{n} \right)^n.$$

For all $n \geq 30$, we then have

$$\frac{10^n}{n!} \leq \frac{1}{2} \left(\frac{10e}{30} \right)^n,$$

so that taking

$$n = \frac{-\ln 10^{-3}}{\ln 3 - 1} = 70.05$$

ought to suffice. For 10^{-6} we need to have $n = 140.1$ These are both conservative estimates, because the ratio $10e/n$ is of course decreasing as n increases, and we took it as the fixed value $r = 10e/30 \approx 0.9061$. Using a program like Maple we can find that the precise values are 32 and 38.

3. Use Stirling's formula to determine the value of

$$\lim_{n \to \infty} \frac{(n!)^p}{(pn)!},$$

where $p \geq 2$ is an integer.

Solution: We have

$$\frac{(n!)^p}{(pn)!} = \frac{[C_n \sqrt{n}(n/e)^n]^p}{C_{pn} \sqrt{pn}(pn/e)^{pn}} = \left(\frac{C_n^p}{C_{pn} \sqrt{p}} \right) \left(\frac{\sqrt{n^{p-1}}}{\sqrt{p}} \right) \left(\frac{1}{p^p} \right)^n = \left(\frac{\sqrt{n^{p-1}}}{C_{pn} \sqrt{p}} \right) \left(\frac{C_n}{p^n} \right)^p.$$

It is a simple exercise with L'Hôpital's rule to show that the second factor goes to zero much faster than the first one goes to infinity. Thus the limit is zero.

4. Use Stirling's formula to get an upper bound on the ratio

$$R_n = \frac{1 \cdot 3 \cdot 5 \cdots (2n-1)}{2^{n+1} n!}$$

as a function of n.

Solution: The idea is to insert factors of 2 into the numerator and denominator so that the product of odd terms can be treated as a factorial:

$$\frac{1 \cdot 3 \cdot 5 \cdots (2n-1)}{2^{n+1} n!} = \frac{1 \cdot 2 \cdot 3 \cdot 4 \cdot 5 \cdots (2n-1) \cdot (2n)}{[2 \cdot 4 \cdot 6 \cdots (2n)] 2^{n+1} n!} = \frac{(2n)!}{2^n (n!) 2^{n+1} n!} = \frac{(2n)!}{2(2^n n!)^2}.$$

Using Stirling's formula we get

$$R_n \leq \frac{C_{2n} \sqrt{2n} (2n/e)^{2n}}{2 c_n^2 n (2n/e)^{2n}} = \frac{C_{2n} \sqrt{2n}}{2 c_n^2 n} = \frac{C_{2n}}{2 c_n^2} n^{-1/2},$$

where $C_k \leq 2.501$ and $c_k \geq 2.37$, for all k. Thus we have

$$R_n \leq \frac{0.2226}{n^{1/2}}.$$

◁ • • • ▷

5.6 GAUSSIAN QUADRATURE

Exercises:

1. Apply Gaussian quadrature with $n = 4$ to approximate each of the following integrals.

  (a)
$$I = \int_{-1}^{1} \ln(1 + x^2) dx = 2 \ln 2 + \pi - 4.$$

  Solution:

$$\begin{aligned}
G_4(f) = {} & (0.3478548451374476) \ln[1 + (-0.8611363115940526)^2] \\
& + (0.6521451548625464) \ln[1 + (-0.3399810435848563)^2] \\
& + (0.6521451548625464) \ln[1 + (0.3399810435848563)^2] \\
& + (0.3478548451374476) \ln[1 + (0.8611363115940526)^2] \\
= {} & 0.5286293488.
\end{aligned}$$

  (b)
$$I = \int_{-1}^{1} \sin^2 \pi x \, dx = 1.$$

Solution:

$$G_4(f) = (0.3478548451374476)\sin^2[\pi(-0.8611363115940526)]$$
$$+ (0.6521451548625464)\sin^2[\pi(-0.3399810435848563)]$$
$$+ (0.6521451548625464)\sin^2[\pi(0.3399810435848563)]$$
$$+ (0.3478548451374476)\sin^2[\pi(0.8611363115940526)]$$
$$= 1.125732289.$$

(c)

$$I = \int_{-1}^{1} (x^8 + 1)dx = 20/9.$$

Solution:

$$G_4(f) = (0.3478548451374476)((-0.8611363115940526)^8 + 1)$$
$$+ (0.6521451548625464)((-0.3399810435848563)^8 + 1)$$
$$+ (0.6521451548625464)((0.3399810435848563)^8 + 1)$$
$$+ (0.3478548451374476)((0.8611363115940526)^8 + 1)$$
$$= 2.210612246.$$

(d)

$$I = \int_{-1}^{1} e^{-x^2}dx = 1.493648266.$$

Solution:

$$G_4(f) = (0.3478548451374476)e^{-(-0.8611363115940526)^2}$$
$$+ (0.6521451548625464)e^{-(-0.3399810435848563)^2}$$
$$+ (0.6521451548625464)e^{-(0.3399810435848563)^2}$$
$$+ (0.3478548451374476)e^{-(0.8611363115940526)^2}$$
$$= 1.493334622.$$

(e)

$$I = \int_{-1}^{1} \frac{1}{1 + x^4}dx = 1.733945974.$$

Solution:

$$G_4(f) = (0.3478548451374476)\frac{1}{1 + (-0.8611363115940526)^4}$$
$$+ (0.6521451548625464)\frac{1}{1 + (-0.3399810435848563)^4}$$
$$+ (0.6521451548625464)\frac{1}{1 + (0.3399810435848563)^4}$$
$$+ (0.3478548451374476)\frac{1}{1 + (0.8611363115940526)^4}$$
$$= 1.735966736.$$

2. Apply Gaussian quadrature with $n = 4$ to approximate each of the following integrals. Remember that you have to do a change of variable to $[-1, 1]$ first.

   (a)
   $$I = \int_0^1 \ln(1 + x)dx = 2\ln 2 - 1.$$

   (b)
   $$I = \int_0^1 \frac{1}{\sqrt{1 + x^4}}dx = 0.92703733865069.$$

   Solution: $G_4(f) = 0.9270388618.$

   (c)
   $$I = \int_0^1 x(1 - x^2)dx = \frac{1}{4}.$$

   (d)
   $$I = \int_0^1 \frac{1}{1 + x^3}dx = \frac{1}{3}\ln 2 + \frac{1}{9}\sqrt{3}\pi.$$

   Solution: $G_4(f) = 0.8356239692.$

   (e)
   $$I = \int_1^2 e^{-x^2}dx = 0.1352572580.$$

3. Show that (5.8) is both necessary and sufficient to make the quadrature exact for all polynomials of degree less than or equal to $2N - 1$.

   Solution: Suppose (5.8) holds, and let $q(x)$ be an arbitrary polynomial of degree $\leq 2N - 1$. Then
   $$q(x) = \sum_{i=0}^{2N-1} a_i x^i$$

   and
   $$\int_{-1}^1 q(x)dx = \sum_{i=0}^{2N-1} a_i \int_{-1}^1 x^i dx.$$

   Since Gaussian quadrature is exact for this range of degree of polynomials, we then have
   $$I(q) = \sum_{i=0}^{2N-1} a_i \left(\sum_{k=1}^{n} w_k^{(n)} [x_k^{(n)}]^i \right) = G_n(q).$$

   The steps are entirely reversible, so we can easily prove the other direction of the implication.

4. Write a program that does Gaussian quadrature, using the weights and Gauss points given in the text. Apply this to each of the integrals below, and compare your results to those for the other quadrature methods in the exercises. Remember that you will have to do a change of variable if the interval is not $[-1, 1]$. Be sure to run your program for more than one value of n.

   (a) $f(x) = \sqrt{1 - x^2}, [-1, 1], I(f) = \pi/2;$

(b) $f(x) = x^2 e^{-x}$, $[0, 2]$, $I(f) = 2 - 10e^{-2} = 0.646647168$;

   Solution: $G_4(f) = 0.6466426358$.

(c) $f(x) = \ln x$, $[1, 3]$, $I(f) = 3\ln 3 - 2 = 1.295836867$;

(d) $f(x) = 1/(1 + x^2)$, $[-5, 5]$, $I(f) = 2\arctan(5)$;

(e) $f(x) = e^{-x}\sin(4x)$, $[0, \pi]$, $I(f) = \frac{4}{17}(1 - e^{-\pi}) = 0.2251261368$.

5. Let $P(x) = 6x^3 + 5x^2 + x$, and let $P_2(x) = 3x^2 - 1$ (this is the quadratic Legendre polynomial). Find linear polynomials $Q(x)$ and $R(x)$ such that $P(x) = P_2(x)Q(x) + R(x)$. Verify that $I(P) = I(R)$.

   Solution: Simple division shows that $Q(x) = 2x + (5/3)$ and $R(x) = 3x + (5/3)$. It is then easily verified that

   $$(3x^2 - 1)(2x + (5/3)) + (3x + (5/3)) = P(x).$$

6. Let $P(x) = x^3 + x^2 + x - 1$, and repeat the above.

   Solution: This time $Q(x) = (1/3)(x + 1)$ and $R(x) = (4/3)x - (2/3)$.

7. Let $P(x) = 3x^3 + x^2 - 6$, and repeat the above.

   Solution: This time $Q(x) = x + (1/3)$ and $R(x) = x - (17/3)$.

8. Verify that the weights for the $n = 2$ Gaussian quadrature rule satisfy the formula (5.9).

   Solution: This amounts to showing that

   $$\int_{-1}^{1} \frac{x - 0.5773502691896257}{-0.5773502691896257 - 0.5773502691896257} dx = 1$$

   which it does, and similarly for the other Gauss point and weight.

9. Repeat the above for the $n = 4$ rule.

   Solution: This involves the same kind of computation, with the same result.

10. Show, by direct computation, that the $n = 2$ and $n = 4$ Gaussian quadrature rules are exact for the correct degree of polynomials.

    Solution: Direct computation does the trick.

11. The quadratic Legendre polynomial is $P_2(x) = (3x^2 - 1)/2$. Show that it is orthogonal (over $[-1, 1]$) to all linear polynomials.

    Solution: Let $L(x) = Ax + B$ be an arbitrary linear polynomial; then we have

    $$\int_{-1}^{1} P_2(x)L(x)dx = \frac{1}{2}\int_{-1}^{1} \left(3Ax^3 + 3Bx^2 - Ax - B\right) dx = 0$$

    for any A and B.

12. The cubic Legendre polynomial is $P_3(x) = (5x^3 - 3x)/2$. Show that it is orthogonal (over $[-1, 1]$) to all polynomials of degree less than or equal to 2.

    Solution: Essentially the same proof as the above.

13. The quartic Legendre polynomial is $P_4(x) = (35x^4 - 30x^2 + 3)/8$. Show that it is orthogonal (over $[-1, 1]$) to all polynomials of degree less than or equal to 3.

    Solution: Essentially the same proof, again, only longer.

14. The first two Legendre polynomials are

$$P_0(x) = 1, \quad P_1(x) = x,$$

and it can be shown that the others satisfy the recurrence relation

$$(n + 1)P_{n+1}(x) = (2n + 1)xP_n(x) - nP_{n-1}(x).$$

Use this to show (by induction) that the leading coefficient for the Legendre polynomials satisfies

$$k_n = \frac{(2n)!}{2^n(n!)^2}$$

and the 2-norm of the Legendre polynomials satisfies

$$\|P_n\|_2^2 = \frac{2}{2n + 1}.$$

Solution: Since $k_0 = 1$ and $k_1 = 1$, the formula works for the first two Legendre polynomials. The recurrence then implies that

$$(n + 1)k_{n+1} = (2n + 1)k_n \Rightarrow k_{n+1} = \left(\frac{2n + 1}{n + 1}\right)\left(\frac{(2n)!}{2^n(n!)^2}\right),$$

$$k_{n+1} = \left(\frac{2n + 2}{2n + 2}\right)\left(\frac{2n + 1}{n + 1}\right)\left(\frac{(2n)!}{2^n(n!)^2}\right) = \frac{(2n + 2)!}{2^{n+1}((n + 1)!)^2}.$$

A similar direct computation establishes the norm result.

15. Let

$$I(f) = \int_a^b f(x)dx.$$

Show that the change of variable $x = a + \frac{1}{2}(b - a)(z + 1)$ gives that

$$I(f) = \int_{-1}^1 F(z)dz$$

for $F(z) = \frac{1}{2}(b - a)f(a + \frac{1}{2}(b - a)(z + 1))$.

Solution: Direct computation.

16. Show that the error for Gaussian quadrature applied to

$$I(f) = \int_a^b f(x)dx$$

is $\mathcal{O}([(b - a)(e/8n)]^{2n})$.

Solution: Apply Stirling's formula to the error estimate.

17. Prove Theorem 5.5. Hint: Simply generalize what was done in the text for the special case of

$$I(f) = \int_{-1}^{1} f(x)dx.$$

Solution: This is a routine adaptation of the exposition in the text to the situation in which the polynomials ϕ_k are orthogonal with respect to the weighted inner product defined by w and $[a, b]$.

18. Once again, we want to consider the approximation of the natural logarithm function, this time using numerical quadrature. Recall that we have

$$\ln x = \int_{1}^{x} \frac{1}{t}dt.$$

Recall also that it suffices to consider $x \in [\frac{1}{2}, 1]$.

(a) How many grid points are required for 10^{-16} accuracy using the trapezoid rule? Simpson's rule?

(b) How many grid points are required if Gauss-Legendre quadrature is used?

Solution: Using the formula for the Gaussian quadrature error over a general interval, we get (assuming x is near 1 to maximize the $b - a$ factor):

$$
\begin{aligned}
|\ln x - G_n| &= \frac{(1/2)^{2n+1}(n!)^4}{(2n+1)[(2n)!]^3} \times \frac{(2n-1)!}{(1/2)^{2n}} = \frac{(1/2)}{(2n+1)(2n)} \times \frac{(n!)^4}{[(2n)!]^2} \\
&\leq \frac{1}{8n^2} \times \frac{(n!)^4}{[(2n)!]^2}.
\end{aligned}
$$

We use Stirling's formula to estimate the factorials. We have

$$(n!)^4 \leq (2.501)^4 n^2 (n/e)^{4n}$$

and

$$[(2n)!]^2 \geq (2.3)^2 (2n)(2n/e)^{4n}.$$

Therefore

$$|\ln x - G_n| \leq \frac{(2.501)^4 n^2}{16(2.3)^2 n^3} \times (1/2)^{4n} = \frac{(0.46)(1/2)^{4n}}{n}.$$

This achieves the desired accuracy for $n \geq 13$.

19. Write a computer program that uses Gaussian quadrature for a specified number of points to compute the natural logarithm over the interval $[\frac{1}{2}, 1]$, to within 10^{-16} accuracy. Compare the accuracy of your routine to the intrinsic logarithm function on your system.

Solution: A MATLAB script is given below. The maximum error on $[1/2, 1]$ was about 4×10^{-15}. This is larger than the specified accuracy, but this is probably due to rounding error within MATLAB itself.

```
function y = loggauss(x)
    a =    [0.9501250983763744E-01    0.1894506104550685E+00
```

$$
\begin{array}{ll}
0.2816035507792589E+00 & 0.1826034150449236E+00 \\
0.4580167776572274E+00 & 0.1691565193950024E+00 \\
0.6178762444026438E+00 & 0.1495959888165733E+00 \\
0.7554044083550030E+00 & 0.1246289712555339E+00 \\
0.8656312023878318E+00 & 0.9515851168249290E-01 \\
0.9445750230732326E+00 & 0.6225352393864778E-01 \\
0.9894009349916499E+00 & 0.2715245941175185E-01];
\end{array}
$$

```
      xg = a(:,1);
      wg = a(:,2);
      xg = [-xg; xg];
          wg = [wg; wg];
%
      xg = 1 + 0.5*(x-1)*(xg + 1);
      wg = 0.5*(x-1)*wg;
%
      y = sum(wg./xg);
```

20. Write a brief essay, in your own words, of course, which explains the importance of the *linearity* of the integral and quadrature rule in the development of Gaussian quadrature.

    Solution: The linearity is vital because of the role played by inner products and orthogonality in the construction and accuracy of the Gaussian quadrature.

◁ • • • ▷

5.7 EXTRAPOLATION METHODS

Exercises:

1. Apply Simpson's rule with $h = \frac{1}{2}$ and $h = \frac{1}{4}$ to approximate the integral

$$
I = \int_0^1 \frac{1}{\sqrt{1+x^4}}dx = 0.92703733865069,
$$

   and use Richardson extrapolation to obtain the improved value of the approximation. What is the *estimated* value of the error in S_4, compared to the actual error?

   Solution:

$$
S_2(f) = 0.9312794637, \quad S_4(f) = 0.9271586870, \quad R_4(f) = 0.9268839688,
$$

$$
E_4(f) = 0.27471844 \times 10^{-3}, \quad I(f) - R_4 = 0.1533703 \times 10^{-3}.
$$

2. Repeat Problem 1, for
$$
I = \int_0^1 x(1 - x^2)dx = \frac{1}{4}.
$$

   Solution: For this exercise, everything is exact.

3. Repeat Problem 1, for

$$I = \int_0^1 \ln(1+x)dx = 2\ln 2 - 1.$$

Solution:

$$S_2(f) = 0.3858346022, \quad S_4(f) = 0.3862595628, \quad R_4(f) = 0.3862878936,$$

$$E_4(f) = -0.2833071 \times 10^{-4}, \quad I(f) - R_4 = 0.6467 \times 10^{-5}.$$

4. Repeat Problem 1, for

$$I = \int_0^1 \frac{1}{1+x^3}dx = \frac{1}{3}\ln 2 + \frac{1}{9}\sqrt{3}\pi.$$

Solution:

$$S_2(f) = 0.8425925926, \quad S_4(f) = 0.8357855108, \quad R_4(f) = 0.8353317056,$$

$$E_4(f) = 0.45380545 \times 10^{-3}, \quad I(f) - R_4 = 0.3171429 \times 10^{-3}.$$

5. Repeat Problem 1, for

$$I = \int_1^2 e^{-x^2}dx = 0.1352572580.$$

Solution:

$$S_2(f) = 0.1346319964, \quad S_4(f) = 0.1352101306, \quad R_4(f) = 0.1352486730,$$

$$E_4(f) = -0.38542280 \times 10^{-5}, \quad I(f) - R_4 = 0.85850 \times 10^{-5}.$$

6. Write a trapezoid rule program to compute the value of the integral

$$I = \int_0^1 e^{-x^2}dx.$$

Take h small enough to justify a claim of accuracy to within 10^{-6}, and explain how the claim is justified. (There are several ways of doing this.)

Solution: There are three primary ways of doing this: (1) use the error estimate to predict how small h must be to theoretically guarantee an error of less than 10^{-6}; (2) use the correction term from the corrected trapezoid rule as an error estimator, and stop when that estimate is less than 10^{-6} in absolute value; (3) use the Richardson estimate of the error , and stop when that estimate is less than 10^{-6} in absolute value. The exercise was written with (3) in mind, but the other two are equally correct.

7. Define

$$I(f) = \int_0^1 e^x \, dx$$

and consider the approximation of this integral using Simpson's rule together with extrapolation. By computing a sequence of approximate values S_2, S_4, S_8, etc., determine experimentally the accuracy of the extrapolated rule

$$R_{2n} = (16S_{2n} - S_n)/15.$$

Solution: You should be able to show that $I(f) - R_{2n} = \mathcal{O}(h^6)$.

8. Consider the integral

$$I(f) = \int_0^\pi \sin^2 x \, dx = \frac{1}{2}\pi.$$

Write a trapezoid rule or Simpson's rule program to approximate this integral, using Richardson extrapolation to improve the approximations, and comment on your results. In particular, comment upon the rate at which the error decreases as h decreases, and on the amount of improvement obtained by extrapolation.

Solution: You shouldn't see much improvement, because you are integrating a periodic function over an integer multiple of the period. We shall see in §5.8 that this kind of computation will be super-accurate. For example, for $n > 2$, Simpson's rule was exact to 14 digits, so the Richardson process didn't really need to be invoked.

9. Repeat the above, this time for the integral

$$I(f) = \int_0^{3\pi/4} \sin^2 x \, dx = \frac{1}{4} + \frac{3}{8}\pi.$$

Solution: Since the interval of integration is not an integer multiple of the period of the integrand, the super-accuracy is lost, and we see the kind of behavior we expect from the usual theory.

10. The data in Table 5.4 (Table 5.11 in the text) supposedly comes from applying the midpoint rule to a smooth, non-periodic function. Can we use this data to determine whether or not the program is working properly? Explain.

Table 5.4   Data for Problem 10.

n	$M_n(f)$
4	−0.91595145
8	−0.95732875
16	−0.97850187
32	−0.98921026
64	−0.99459496
128	−0.99729494
256	−0.99864683
512	−0.99932326
1024	−0.99966159

Solution: If we compute the estimated exponent p, we get that $p \approx 1$, but for the midpoint rule we expect $p = 2$, so we conclude that the program has an error.

11. The error function, defined as

$$\text{erf}(x) = \frac{2}{\sqrt{\pi}} \int_0^x e^{-t^2} dt$$

is an important function in probability theory and heat conduction. Use Simpson's rule with extrapolation to produce values of the error function that are accurate to within 10^{-8} for x ranging from 0 to 5 in increments of $1/4$. Check your values against the intrinsic error function on your computer or by looking at a set of mathematical tables such as *Handbook of Mathematical Functions*.

Solution: For the range of values of interest, the extrapolated estimate is sufficiently accurate with at most $n = 64$ subintervals.

12. Bessel functions appear in the solution of heat conduction problems in a circular or cylindrical geometry, and can be defined as a definite integral, thus:

$$J_k(x) = \frac{1}{\pi} \int_0^\pi \cos(x \sin t - kt) dt.$$

Use Simpson's rule plus extrapolation to produce values of J_0 and J_1 that are accurate to within 10^{-8} over the interval $[0, 6]$ in increments of $1/4$. Check your values by looking at a set of tables such as *Handbook of Mathematical Functions*.

Solution: Again, taking $n = 64$ yields sufficient accuracy for the extrapolated value over this range of arguments. Sometimes we need only $n = 16$.

13. Apply the trapezoid rule to approximate the integral

$$I = \int_0^{\pi^2/4} \sin \sqrt{x} dx = 2.$$

Use Richardson extrapolation to estimate how rapidly your approximations are converging, and comment on your results in the light of the theory for the trapezoid rule.

Solution: The estimated exponent p goes to 1.5 instead of the expected 2. In other words, the approximation is not as accurate as it should be. This occurs because of the square root in the integrand.

14. Show that, for any function f,

$$S_2(f) = (4T_2(f) - T_1(f))/3.$$

Comment on the significance of this result in light of this section.

Solution: The result is proved by a direct computation. We have

$$T_1(f) = \frac{b - a}{2}(f(a) + f(b))$$

and

$$T_2(f) = \frac{(b - a)/2}{2}(f(a) + 2f((a + b)/2) + f(b)).$$

Simply add these up, as indicated, to get $S_2(f)$.

The significance of this is that it shows that Richardson extrapolation applied to the trapezoid rule yields Simpson's rule.

◁ ● ● ● ▷

5.8 SPECIAL TOPICS IN NUMERICAL INTEGRATION

Exercises:

1. Use the Euler-Maclaurin formula to state and prove a formal theorem that the corrected trapezoid rule, T_n^C, is $\mathcal{O}(h^4)$ accurate.

   Solution: For $N = 1$, the Euler-Maclaurin formula says

   $$I(f) - T_n(f) = -\frac{1}{12}h^2(f'(b) - f'(a)) + \frac{(b-a)}{720}h^4 f^{(4)}(\xi), \quad (5.1)$$

   from which

   $$I(f) - T_n^C(f) = \mathcal{O}(h^4)$$

   follows immediately.

2. Using a hand calculator (or calculator app), compute $T_1^{(0)}(f), T_2^{(0)}(f), T_4^{(0)}(f)$, and $T_8^{(0)}(f)$ for each of the following functions, then use Romberg integration to compute $\Theta_3(f)$. Note: Be sure to use Theorem 5.8 to minimize the work in computing the first column of the Romberg array.

   (a)
   $$I = \int_0^1 \ln(1+x)dx = 2\ln 2 - 1.$$

   (b)
   $$I = \int_0^1 \frac{1}{\sqrt{1+x^4}}dx = 0.92703733865069.$$

   Solution: The Romberg array is

   0.8535533905
   0.9118479454   0.9312794632
   0.9233310016   0.9271586869   0.9268839688
   0.9261151802   0.9270432395   0.9270355433   0.9270379494

   (c)
   $$I = \int_0^1 x(1-x^2)dx = \frac{1}{4}.$$

   (d)
   $$I = \int_0^1 \frac{1}{1+x^3}dx = \frac{1}{3}\ln 2 + \frac{1}{9}\sqrt{3}\pi.$$

Solution: $\Theta_3(f) = 0.8356560744.$

(e)

$$I = \int_1^2 e^{-x^2} dx = 0.1352572580.$$

Solution: $\Theta_3(f) = 0.1352573547.$

3. Write a program to do Romberg integration. Be sure to use Theorem 5.8 in order to minimize the number of function evaluations. Test your program by applying it to the following example problems.

   (a) $f(x) = \ln x$, $[1, 3]$, $I(f) = 3\ln 3 - 2 = 1.295836867$;
   (b) $f(x) = x^2 e^{-x}$, $[0, 2]$, $I(f) = 2 - 10e^{-2} = 0.646647168$;
   (c) $f(x) = \sqrt{1 - x^2}$, $[-1, 1]$, $I(f) = \pi/2$;
   (d) $f(x) = 1/(1 + x^2)$, $[-5, 5]$, $I(f) = 2\arctan(5)$;
   (e) $f(x) = e^{-x}\sin(4x)$, $[0, \pi]$, $I(f) = \frac{4}{17}(1 - e^{-\pi}) = 0.2251261368.$

   For each example, compute

   $$N_f = \frac{\text{Number of function evaluations}}{-\log_{10}|\text{error}|}.$$

   This measures the number of function evaluations needed to produce each correct decimal digit in the approximation.

   Solution: A MATLAB script for Romberg integration is given below. It is easily modified to compute N_f, which can behave erratically for some of these examples.

```
function r = romberg(m,a,b)
    Tnew = zeros(1,m+1);
    Told = Tnew;
    T0 = 0.5*(b-a)*(romf(b) + romf(a));
    Told(1) = T0;
    for k=1:m
        n = 2^k;
        hn = (b-a)/n;
        x = a + [1:2:(n-1)]*hn;
        y = romf(x);
        T0 = Told(1);
        T1 = 0.5*T0 + hn*sum(y);
        Tnew(1) = T1;
        for j=1:k
            R = (4^j*Tnew(j) - Told(j))/(4^j - 1);
            Tnew(j+1) = R;
        end
        Told = Tnew;
    end
    r = Tnew(m+1);
```

4. Write a computer program that uses Romberg integration for a specified number of points to compute the natural logarithm over the interval $[\frac{1}{2}, 1]$, to within 10^{-16}

accuracy. Compare the accuracy of your routine to the intrinsic logarithm function on your system.

Solution: The romberg script can be easily modified to do this. The trick is to use the Richardson error estimate to terminate the computation when the specified accuracy is achieved.

5. Show that the change of variable $x = \phi(t)$, where ϕ is as given in §5.8.2, transforms the integral

$$I(f) = \int_a^b f(x)dx$$

into the integral

$$I(f) = \int_{-1}^1 f(\phi(t))\phi'(t)dt.$$

Solution: This is a straight-forward calculus exercise.

6. Apply the singular integral technique of §5.8.2, with $n = 4$, to estimate the value of each of the following integrals. Do this with a hand calculator, using the values in Table 5.14 (in the text).

(a)
$$I(f) = \int_0^1 \frac{\ln x}{1-x^2}dx = -\frac{\pi^2}{8}.$$

   Solution: $I_4 = -1.28351724985458.$

(b)
$$I(f) = \int_0^1 \frac{\ln x}{1-x}dx = -\frac{\pi^2}{6};$$

   Solution: $I_4 = -1.68916522566720.$

(c)
$$I(f) = \int_0^1 x\ln(1-x)dx = -\frac{3}{4};$$

   Solution: $I_4 = -0.79940423863966.$

(d)
$$I(f) = \int_0^1 \left(\ln\frac{1}{x}\right)^{-1/2} dx = \sqrt{\pi};$$

   Solution: $I_4 = 1.83057018789044.$

   None of these is especially accurate, but recall we are only using $n = 4$.

7. Repeat the previous problem, except this time use a computer program together with the values in Table 5.14 (from the text) to compute the $n = 16$ approximations.

   (a) **Solution:** $I_{16} = -1.23368545936076.$

   (b) **Solution:** $I_{16} = -1.64491714791208.$

   (c) **Solution:** $I_{16} = -0.74998668656848.$

   (d) **Solution:** $I_{16} = 1.77212889191835.$

These are substantial improvements over the $n = 4$ values.

8. We have looked at the gamma function in a number of exercises in previous chapters. The formal definition of $\Gamma(x)$ is the following:

$$\Gamma(x) = \int_0^\infty e^{-t} t^{x-1} dt.$$

Use the infinite interval algorithm from §5.8.2 to construct a table of values for the gamma function over the interval $[1, 2]$. Compare your results to the values you get for $\Gamma(x)$ on your computer or from a standard book of tables.

Solution: Using the $n = 16$ version gives a maximum error of about 10^{-3}.

9. Modify your Romberg integration program to compute values of $\Psi(t_k)/\Psi(1)$ for $t_k \in [-1, 1]$, and use this to extend the values in Table 5.14 to the $n = 64$ case.

Solution: A partial table of $\Psi(t_k)/\Psi(1)$ values for $n = 64$ is in Table 5.5.

Table 5.5 Table of $\Psi(t_k)/\Psi(1)$ for $-1 \le t_k \le 1$.

t_k	$\Psi(t_k)/\Psi(1)$
−0.250000	0.20043174541744
−0.218750	0.23268738852437
−0.187500	0.26694920180823
−0.156250	0.30298089532632
−0.125000	0.34052262810721
−0.093750	0.37929519918613
−0.062500	0.41900410866588
−0.031250	0.45934349927578
0.0	0.50000000000000
0.031250	0.54065650072422
0.062500	0.58099589133412
0.093750	0.62070480081387
0.125000	0.65947737189279
0.156250	0.69701910467368
0.187500	0.73305079819177
0.218750	0.76731261147563
0.250000	0.79956825458256

10. Apply the trapezoid rule to each of the following functions, integrated over the indicated intervals, and interpret the results in terms of the Euler-Maclaurin formula.

   (a) $f(x) = 1 + \sin \pi x$, $[a, b] = [0, 2]$;
   (b) $f(x) = \sin^2 x$, $[a, b] = [0, \pi]$.

Solution: In both cases, the trapezoid rule produces the *exact* value for $n = 2$, which is remarkable accuracy. This is because we are integrating periodic functions over an interval that is an integer multiple of the period. The Euler-Maclaurin formula predicts that this will result in exceptional accuracy, and it does.

11. Using a hand calculator (or calculator app) and τ as indicated, perform the adaptive quadrature scheme outlined in §5.8.3 on each of the following integrals. Be sure to present your results in an orderly fashion so that the progress of the calculation can be followed.

(a) $\tau = 5 \times 10^{-6}$;

$$I = \int_0^1 \ln(1 + x)dx = 2\ln 2 - 1.$$

(b) $\tau = 10^{-5}$;

$$I = \int_0^1 \frac{1}{1 + x^3}dx = \frac{1}{3}\ln 2 + \frac{1}{9}\sqrt{3}\pi.$$

(c) $\tau = 10^{-5}$;

$$I = \int_1^2 e^{-x^2}dx = 0.1352572580.$$

(d) $\tau = 10^{-4}$;

$$I = \int_0^1 \frac{1}{\sqrt{1 + x^4}}dx = 0.92703733865069.$$

Solution: In the first step we get

$$S_1 = 0.9312794637, \quad S_2 = 0.9271586870, \quad R = 0.9268839688,$$
$$E = 0.2747184467 \times 10^{-3}.$$

Since $E > \tau$ we do not accept the value. In the second step, we compute

$$S_1 = 0.4302959524, \quad S_2 = 0.4300938322, \quad R = 0.4300803577,$$
$$E = 0.1347468000 \times 10^{-4}.$$

This time, we have $E \le \tau/2$ so we accept R as the partial integral of f to the specified accuracy:

$$\int_{1/2}^1 f(x)dx = 0.4300803577.$$

We then continue the process, trying to approximate the rest of the integral.

12. Apply the MATLAB routines quad, quadl, and quadgk to each of the following integrations, with a tolerance $\tau = 10^{-8}$ in each case, then repeat the computations over the left half of the interval, only:

(a)

$$I = \int_0^1 \frac{1}{1 + 1023e^{-16t}}dt = 0.56679020695363;$$

(b)

$$I = \int_0^{2\pi} e^{\sin 4\pi x}dx = 8.11767960946423;$$

(c)

$$I = \int_0^1 \sin(e^{\pi x})dx = 0.20499307668744;$$

(d)

$$I = \int_0^1 \frac{1}{1+x^3} dx = \frac{1}{3} \ln 2 + \frac{1}{9} \sqrt{3}\pi;$$

(e)

$$I = \int_0^1 \frac{1}{\sqrt{1+x^4}} dx = 0.92703733865069.$$

13. This is an experimental or research problem. Try to find a specific quadrature problem

$$I = \int_a^b f(x) dx$$

such that quad outperforms quadl consistently as the tolerance τ decreases. Then try to find a different one such that quadl outperforms quad.

14. Show that the trapezoid rule $T_1(f)$ is exact for all functions of the form $f(x) = Ax + B$.

Solution: If $f(x) = Ax + B$, then

$$T_1(f) = \frac{b-a}{2}(Aa + B + Ab + B) = \frac{1}{2}A(b^2 - a^2) + B(b - a) = I(f).$$

15. Show that

$$I(\alpha f + \beta g) = \alpha I(f) + \beta I(g)$$

for constants α and β, and similarly for $T_n(\alpha f + \beta g)$.

Solution:

$$\begin{aligned} I(\alpha f + \beta g) &= \int_a^b (\alpha f(x) + \beta g(x)) dx = \int_a^b \alpha f(x) dx + \int_a^b \beta g(x) dx \\ &= \alpha \int_a^b f(x) dx + \beta \int_a^b g(x) dx = \alpha I(f) + \beta I(g), \end{aligned}$$

and the same argument holds for T_n, of course.

16. Show that the data in Table 5.13 (in the text) confirms that the trapezoid and Simpson's rules applied to $f(x) = \sqrt{x}$ are both $\mathcal{O}(h)$ accurate.

Solution: Using Richardson extrapolation we get that (for both rules) the exponent $p \approx 1.5$ which suggests that both rules are actually $\mathcal{O}(h^{3/2})$.

17. Confirm that (5.28) is the correct Peano kernel for the composite trapezoid rule.

Solution: Write the integral as

$$I(f) = \sum_{k=1}^n \int_{t_{k-1}}^{t_k} f(t) dt$$

and apply the Peano kernel to each integral over $[k-1, t_k]$.

18. Show that if f' is integrable over $[a, b]$, but f'' is not, then the trapezoid rule is $\mathcal{O}(h)$ accurate.

Solution: We have

$$I(f) - T_n(f) = \int_a^b K(t) f''(t) dt.$$

Integration by parts can be carefully used to rewrite this as

$$I(f) - T_n(f) = -\int_a^b K'(t) f'(t) dt.$$

(The additional terms from integration by parts involve $K(x_i)$ and therefore vanish.) Hence,

$$|I(f) - T_n(f)| \leq \max_{a \leq t \leq b} |K'(t)| \int_a^b |f'(t)| dt.$$

19. Show that the Peano Theorem implies an error estimate for the trapezoid rule of the form

$$|I(f) - T_n(f)| \quad \leq \quad \frac{1}{8} h^2 \int_a^b |f''(t)| dt.$$

Be sure to provide all details missing from the development in the text.

Solution: The only detail left out of the discussion in the text is to show that

$$\max_{a \leq t \leq b} |K(t)| \leq \frac{1}{8} h^2,$$

but this is done the same way as the linear interpolation error estimate back in Chapter 2.

20. Derive the Peano kernel for Simpson's rule.

Solution: Following the development in the text, we get

$$K(t) = \left\{ \begin{array}{ll} (b-t)^4 - \frac{c-a}{18} (b-t)^3 - \frac{b-a}{9} (c-t)^3 & a \leq t \leq c \\ \frac{1}{24} (b-t)^4 - \frac{c-a}{18} (b-t)^3 & c \leq t \leq b \end{array} \right\}$$

CHAPTER 6

NUMERICAL METHODS FOR ORDINARY DIFFERENTIAL EQUATIONS

6.1 THE INITIAL VALUE PROBLEM—BACKGROUND

Exercises:

1. For each initial value problem below, verify (by direct substitution) that the given function y solves the problem.

   (a) $y' + 4y = 0$, $y(0) = 3$; $y(t) = 3e^{-4t}$.

   (b) $y' = t^2/y$, $y(0) = 1$; $y(t) = \sqrt{1 + \frac{2}{3}t^3}$.

   (c) $ty' - y = t^2$, $y(1) = 4$; $y(t) = 3t + t^2$.

   Solution: This is simply a matter of ordinary computation. For (a), we have, for instance,
   $$\frac{d}{dt}\left(3e^{-4t}\right) + 4\left(3e^{-4t}\right) = -12e^{-4t} + 12e^{-4t} = 0,$$
   and $y(0) = 3e^0 = 3$, therefore the given solution is correct.

2. For each initial value problem in Problem 1, write down the definition of f.

   Solution: The function f is simply the right side of the differential equation when put into the standard form used in the text. Thus, for (a), $f(t, y) = -4y$, for (b) $f(t, y) = (y + t^2)/t$, and for (c), $f(t, y) = t^2/y$.

Solutions Manual to Accompany An Introduction to Numerical Methods and Analysis, Third Edition.
James F. Epperson.
© 2021 John Wiley & Sons, Inc. Published 2021 by John Wiley & Sons, Inc.   **185**

3. For each scalar equation below, write out explicitly the corresponding first-order system. What is the definition of f?

(a) $y'' + 9y = e^{-t}$.

(b) $y'''' + y = 1$.

(c) $y'' + \sin y = 0$.

Solution: For (a), we have $w(t) = (y(t), y'(t))^T$ and the system is

$$\frac{d}{dt}\begin{pmatrix} w_1(t) \\ w_2(t) \end{pmatrix} = \frac{d}{dt}\begin{pmatrix} y(t) \\ y'(t) \end{pmatrix} = \begin{pmatrix} y'(t) \\ e^{-t} - 9y(t) \end{pmatrix} = \begin{pmatrix} w_2(t) \\ e^{-t} - 9w_1(t) \end{pmatrix},$$

therefore

$$f(t, w) = \begin{pmatrix} w_2(t) \\ e^{-t} - 9w_1(t) \end{pmatrix}.$$

Similarly, for (b) we get

$$\frac{d}{dt}\begin{pmatrix} w_1(t) \\ w_2(t) \\ w_3(t) \\ w_4(t) \end{pmatrix} = \frac{d}{dt}\begin{pmatrix} y(t) \\ y'(t) \\ y''(t) \\ y'''(t) \end{pmatrix} = \begin{pmatrix} y'(t) \\ y''(t) \\ y'''(t) \\ 1 - w_1(t) \end{pmatrix} = \begin{pmatrix} w_2(t) \\ w_3(t) \\ w_4(t) \\ 1 - w_1(t) \end{pmatrix},$$

therefore

$$f(t, w) = \begin{pmatrix} w_2(t) \\ w_3(t) \\ w_4(t) \\ 1 - w_1(t) \end{pmatrix}.$$

Finally, for (c) we get

$$\frac{d}{dt}\begin{pmatrix} w_1(t) \\ w_2(t) \end{pmatrix} = \frac{d}{dt}\begin{pmatrix} y(t) \\ y'(t) \end{pmatrix} = \begin{pmatrix} y'(t) \\ -\sin y(t) \end{pmatrix} = \begin{pmatrix} w_2(t) \\ -\sin w_1(t) \end{pmatrix},$$

therefore

$$f(t, w) = \begin{pmatrix} w_2(t) \\ -\sin w_1(t) \end{pmatrix}.$$

4. For each initial value problem below, verify (by direct substitution) that the given function y solves the problem.

(a) $y'' + 4y' + 4y = 0$, $y(0) = 0$, $y'(0) = 1$; $y(t) = te^{-2t}$.

(b) $t^2 y'' + 6ty' + 6y = 0$, $y(1) = 1$, $y'(1) = -3$; $y(t) = t^{-3}$.

(c) $y'' + 5y' + 6y = 0$, $y(0) = 1$, $y'(0) = -2$; $y(t) = e^{-2t}$.

Solution: This is a straight-forward calculus exercise.

5. For each initial value problem below, determine the Lipschitz constant, K, for the given rectangle.

(a) $y' = 1 - 3y$, $y(0) = 1$, $R = (-1, 1) \times (0, 2)$;

(b) $y' = y(1 - y)$, $y(0) = \frac{1}{2}$, $R = (-1, 1) \times (0, 2)$;

(c) $y' = y^2$, $y(0) = 1$, $R = (-1, 1) \times (0, 2)$.

Solution: In general, we get the Lipschitz constant by computing f_y and maximizing it over the rectangle of interest. Therefore,

(a) $f_y(t, y) = -3$ for all (t, y), hence $K = 3$;

(b) $f_y(t, y) = 1 - 2y$ for all t and all $y \in (0, 2)$, hence $K = 3$;

(c) $f_y(t, y) = 2y$, hence $K = 4$.

6. Are any of the initial value problems in the previous exercise smooth and uniformly monotone decreasing over the given rectangle? If so, determine the values of M and m.

Solution: The ODE in (a) is smooth and uniformly monotone decreasing because $f_y < 0$ for all (t, y) of interest. In fact, since $f_y(t, y) = -3$ for all (t, y), we have $m = M = 3$.

6.2 EULER'S METHOD

Exercises:

1. Use Euler's method with $h = \frac{1}{4}$ to compute approximate values of $y(1)$ for each of the following initial value problems. Don't write a computer program, use a hand calculator (or calculator app) to produce an orderly table of (t_k, y_k) pairs.

(a) $y' = y(1 - y)$, $y(0) = \frac{1}{2}$;

Solution:

$$
\begin{aligned}
y_1 &= y_0 + hf(t_0, y_0) = (1/2) + (1/4)y(1 - y) = 0.5625, \\
y_2 &= y_1 + hf(t_1, y_1) = 0.6240, \\
y_3 &= y_2 + hf(t_2, y_2) = 0.6827, \\
y_4 &= y_3 + hf(t_3, y_3) = 0.7368.
\end{aligned}
$$

(b) $ty' = y(\sin t)$, $y(0) = 2$.

Solution: We have to use the limiting value of 1 for $\frac{\sin t}{t}$ at $t = 0$:

$$
\begin{aligned}
y_1 &= y_0 + hf(t_0, y_0) = 2 + (1/4)y = 2.5, \\
y_2 &= y_1 + hf(t_1, y_1) = 3.1185, \\
y_3 &= y_2 + hf(t_2, y_2) = 3.8661, \\
y_4 &= y_3 + hf(t_3, y_3) = 4.7445.
\end{aligned}
$$

(c) $y' = y(1 + e^{2t})$, $y(0) = 1$;

Solution:

$$
\begin{aligned}
y_1 &= y_0 + hf(t_0, y_0) = 1 + (1/4)y(e + e^{2t}) = 1.5, \\
y_2 &= y_1 + hf(t_1, y_1) = 2.4933, \\
y_3 &= y_2 + hf(t_2, y_2) = 4.8109, \\
y_4 &= y_3 + hf(t_3, y_3) = 11.404.
\end{aligned}
$$

(d) $y' + 2y = 1$, $y(0) = 2$;

Solution:

$$
\begin{aligned}
y_1 &= y_0 + hf(t_0, y_0) = 2 + (1/4)(1 - 2y) = 1.25, \\
y_2 &= y_1 + hf(t_1, y_1) = 0.875, \\
y_3 &= y_2 + hf(t_2, y_2) = 0.6875, \\
y_4 &= y_3 + hf(t_3, y_3) = 0.59375.
\end{aligned}
$$

2. For each initial value problem above, use the differential equation to produce approximate values of y' at each of the grid points, t_k, $k = 0, 1, 2, 3, 4$.

Solution:

(a) $y'(t_0) = f(t_0, y_0) = 0.25$, $y'(t_1) = 0.2461$, $y'(t_2) = 0.2346$, $y'(t_3) = 0.2166$, $y'(t_4) = 0.1939$.

(b) $y'(t_0) = f(t_0, y_0) = 2$, $y'(t_1) = 2.474$, $y'(t_2) = 2.990$, $y'(t_3) = 3.514$, $y'(t_4) = 3.992$.

(c) $y'(t_0) = f(t_0, y_0) = 2$, $y'(t_1) = 3.973$, $y'(t_2) = 9.271$, $y'(t_3) = 26.37$, $y'(t_4) = 95.67$.

(d) $y'(t_0) = f(t_0, y_0) = -3$, $y'(t_1) = -1.5$, $y'(t_2) = -0.75$, $y'(t_3) = -0.375$, $y'(t_4) = -0.1875$.

3. Write a computer program that solves each of the initial value problems in Problem 1, using Euler's method and $h = 1/16$.

Solution: Below is a MATLAB script that does this for one specific example (1c).

```
function [tc,yc] = euler1(h,n,y0)
    tc = [0];
    yc = [y0];
    yn = y0;
    tn = 0;
    for k=1:n
    y = yn + h*yn*(1 + exp(2*tn));
    t = k*h;
    tc = [tc t];
    yc = [yc y];
    yn = y;
    tn = t;
    end
    tc = tc';
    yc = yc'
    exact = exp(-0.5)*exp(tc + 0.5*exp(2*tc));
%
    figure(1)
    plot(tc,yc)
    figure(2)
    plot(tc,exact - yc)
```

4. For each initial value problem below, approximate the solution using Euler's method using a sequence of decreasing grids $h^{-1} = 2, 4, 8, \ldots$. For those problems where

an exact solution is given, compare the accuracy achieved over the interval $[0, 1]$ with the theoretical accuracy.

(a) $y' + 4y = 1$, $y(0) = 1$; $y(t) = \frac{1}{4}(3e^{-4t} + 1)$.

(b) $y' = -y \ln y$, $y(0) = 3$; $y(t) = e^{(\ln 3)e^{-t}}$.

(c) $y' + y = \sin 4\pi t$, $y(0) = \frac{1}{2}$.

(d) $y' + \sin y = 0$, $y(0) = 1$.

Solution: By making a slight modification to the `euler1` script (above) we can easily solve the two examples for which exact solutions are given, producing the results shown in Table 6.1.

Table 6.1   Solutions to Exercise 6.2.4.

h^{-1}	Max. error for (b)	Max. error for (a)
2	0.8515	0.5950
4	0.2759	0.1940
8	0.0884	0.0846
16	0.0386	0.0399
32	0.0182	0.0194
64	0.0089	0.0096
128	0.0044	0.0048

◁ ● ● ● ▷

6.3 ANALYSIS OF EULER'S METHOD

Exercises:

1. Use Taylor's Theorem to prove that

$$(1 + x)^n \le e^{nx}$$

for all $x > -1$. Hint: expand e^x in a Taylor series, throw away the unnecessary terms, and then take powers of both sides.

Solution: We have

$$e^x = 1 + x + \frac{1}{2}x^2 e^\xi,$$

therefore

$$1 + x \le e^x - \frac{1}{2}x^2 e^\xi < e^x,$$

therefore

$$(1 + x)^n \le e^{nx}.$$

2. For each initial value problem below, use the error theorems of this section to estimate the value of

$$E(h) = \frac{\max_{t_k \leq 1} |y(t_k) - y_k|}{\|y''\|_{\infty,[0,1]}}$$

using $h = 1/16$, assuming Euler's method was used to approximate the solution.

(a) $y' + 4y = 1$, $y(0) = 1$;
Solution: $E(h) \leq 6.70h$.

(b) $y' = y(1 - y)$, $y(0) = 1/2$;
Solution: We have

$$E(h) = \frac{e^{K-1}}{2K}h,$$

where K is the Lipschitz constant for $f(t, y) = y(1 - y)$. We have $K = \max|f_y| = \max|1 - 2y|$, so we need to know the solution to actually complete the estimation. We find $y = (1 + e^{-t})^{-1}$, so $K = 0.46$, which implies that

$$E(h) = 0.635h.$$

(c) $y' = \sin y$, $y(0) = \frac{1}{2}\pi$.

3. Consider the initial value problem

$$y' = e^{-t} - 16y, \quad y(0) = 1.$$

(a) Confirm that this is smooth and uniformly monotone decreasing in y. What is M? What is m?
Solution: Since $f(t, y) = e^{-t} - 16y$ and therefore $f_y(t, y) = -16$ the ODE is indeed smooth and uniformly monotone decreasing in y, with $M = m = 16$.

(b) Approximate the solution using Euler's method and $h = \frac{1}{8}$. Do we get the expected behavior from the approximation? Explain.
Solution: Since this value of h does *not* satisfy the condition $h < M^{-1}$ we do not expect to get the kind of accuracy described in the theorem.

6.4   VARIANTS OF EULER'S METHOD

Exercises:

1. Using the approximation

$$y'(t_n) \approx \frac{y(t_{n+1}) - y(t_{n-1})}{2h},$$

derive the numerical method (6.20) for solving initial value problems. What is the residual? What is the truncation error? Is it a consistent method?

Solution: We have

$$y'(t) = \frac{y(t+h) - y(t-h)}{2h} - \frac{1}{6}h^2 y'''(\theta_{t,h})$$

so that

$$\frac{y(t+h) - y(t-h)}{2h} - \frac{1}{6}h^2 y'''(\theta_{t,h}) = f(t, y(t))$$

or

$$y(t+h) = y(t-h) + 2hf(t, y(t)) + \frac{1}{3}h^3 y'''(\theta_{t,h}).$$

The residual is

$$R_n = \frac{1}{3}h^3 y'''(\theta_{t,h})$$

and the truncation error is

$$T_n = \frac{1}{3}h^2 y'''(\theta_{t,h}).$$

The method is consistent, so long as y''' is continuous.

2. Using the approximation

$$y'(t_{n-1}) \approx \frac{-y(t_{n+1}) + 4y(t_n) - 3y(t_{n-1})}{2h}$$

derive the numerical method (6.21) for solving initial value problems. What is the residual? What is the truncation error? Is it a consistent method?

Solution: The method is consistent, with truncation error $(h^2/3)y'''(\xi)$.

3. Using the approximation

$$y'(t_{n+1}) \approx \frac{3y(t_{n+1}) - 4y(t_n) + y(t_{n-1})}{2h},$$

derive the numerical method (6.22) for solving initial value problems. What is the residual? What is the truncation error? Is it a consistent method?

Solution: Using the material in §4.5 we have

$$y'(t_{n+1}) = \frac{3y(t_{n+1}) - 4y(t_n) + y(t_{n-1})}{2h} + \frac{1}{3}h^2 y'''(\xi_n),$$

so that the differential equation becomes

$$\frac{3y(t_{n+1}) - 4y(t_n) + y(t_{n-1})}{2h} + \frac{1}{3}h^2 y'''(\xi_n) = f(t_{n+1}, y(t_{n+1})),$$

or

$$3y(t_{n+1}) - 4y(t_n) + y(t_{n-1}) = 2hf(t_{n+1}, y(t_{n+1})) - \frac{2}{3}h^3 y'''(\xi_n).$$

Thus,

$$y(t_{n+1}) = \frac{4}{3}y(t_n) - \frac{1}{3}y(t_{n-1}) + \frac{2}{3}hf(t_{n+1}, y(t_{n+1})) - \frac{2}{9}h^3 y'''(\xi_n).$$

The residual is

$$R_n = -\frac{2}{9}h^3 y'''(\xi_n)$$

and the truncation error is

$$T_n = -\frac{2}{9}h^2 y'''(\xi_n).$$

The method is consistent so long as y''' is continuous.

4. Use the trapezoid rule predictor-corrector with $h = \frac{1}{4}$ to compute approximate values of $y(1)$ for each of the following initial value problems. Don't write a computer program, use a hand calculator (or calculator app) to produce an orderly table of (t_k, y_k) pairs.

(a) $y' = y(1 + e^{2t})$, $y(0) = 1$;
   Solution: $y(1) \approx 42.71053582$.

(b) $y' + 2y = 1$, $y(0) = 2$;

(c) $y' = y(1 - y)$, $y(0) = \frac{1}{2}$;
   Solution: $y(1) \approx 0.7303809855$.

(d) $ty' = y(\sin t)$, $y(0) = 2$.

5. Repeat the above, using the method (6.23) as a predictor-corrector, with Euler's method as the predictor. Also use Euler's method to produce the starting value, y_1.

   Solution: For (b), $y(1) \approx 0.7299429169$; for (c), $y(1) \approx 42.82307269$.

6. Write a computer program that solves each of the initial value problems in Problem 4, using the trapezoid rule predictor-corrector and $h = 1/16$.

   Solution: A MATLAB script which is slightly more general than required is given below.

```
function [tc,yc] = trappc(h,n,y0)
    tc = [0];
    yc = [y0];
    yn = y0;
    tn = 0;
    for k=1:n
    yp = yn + h*yn*(1-yn);
    y  = yn + 0.5*h*(yp*(1-yp) + yn*(1-yn));
    t = k*h;
    tc = [tc t];
    yc = [yc y];
    yn = y;
    tn = t;
    end
    tc = tc';
    yc = yc'
    exact = 1./(1 + exp(-tc));
    figure(1)
    plot(tc,yc)
    figure(2)
    plot(tc,exact - yc)
```

7. For each initial value problem below, approximate the solution using the trapezoid rule predictor-corrector with a sequence of decreasing grids $h^{-1} = 2, 4, 8, \ldots$. For those problems where an exact solution is given, compare the accuracy achieved over the interval $[0, 1]$ with the theoretical accuracy.

   (a) $y' + y = \sin 4\pi t$, $y(0) = \frac{1}{2}$.

   (b) $y' + \sin y = 0$, $y(0) = 1$.

   (c) $y' + 4y = 1$, $y(0) = 1$; $y(t) = \frac{1}{4}(3e^{-4t} + 1)$.

   (d) $y' = -y \ln y$, $y(0) = 3$; $y(t) = e^{(\ln 3)e^{-t}}$.

   Solution: The previous MATLAB script can be easily modified to do this.

8. In this problem, we will consider a tumor growth model based on some work of H.P. Greenspan in *J. Theor. Biology*, vol. 56 (1976), pp. 229–242. The differential equation is

   $$R'(t) = -\frac{1}{3}S_i R + \frac{2\lambda\sigma}{\mu R + \sqrt{\mu^2 R^2 + 4\sigma}}, \quad R(0) = a.$$

   Here $R(t)$ is the radius of the tumor (assumed spherical), λ and μ are scale parameters, both $\mathcal{O}(1)$, S_i measures the rate at which cells at the core of the tumor die, and σ is a nutrient level. Take $\lambda = \mu = 1$, $a = 0.25$, $S_i = 0.8$, and $\sigma = 0.25$. Use the trapezoid rule predictor corrector to solve the differential equation, using $h = 1/16$, and show that the tumor radius approaches a limiting value as $t \to \infty$.

   Solution: By about $t = 5$, it is apparent that the limiting radius is about 0.416.

9. Repeat the previous problem, but this time with $S_i = 0.90$, $\sigma = 0.05$, and $a = 0.50$. What happens now?

   Solution: The limiting size of the tumor is about 0.171.

10. Now solve the differential equation using a variety of S_i, σ, and a values (your choices). What happens?

    Solution: For S_i sufficiently small (or σ sufficiently large) the tumor *grows* in size, instead of shrinking.

11. Now let's model some treatment for our mathematical tumor. In the previous three problems, we assumed that the nutrient level was constant. Suppose we are able to decrease the nutrient level according to the model

    $$\sigma(t) = \sigma_\infty + (\sigma_0 - \sigma_\infty)e^{-qt}.$$

    Here σ_0 is the initial nutrient level, σ_∞ is the asymptotic nutrient level, and q measures the rate at which the nutrient level drops. Investigate the effect of various choices of these parameters on the growth of the tumor, based on your observations from the previous problems. Again, use the trapezoid rule predictor corrector with $h = 1/16$ to solve the differential equation.

    Solution: Generally, the same kind of results occur. The faster and farther the nutrient level is reduced, the more the tumor shrinks.

12. Verify by direct substitution that (6.43) satisfies the recursion (6.39) for all $n \geq 2$.

Solution: Straight forward computation.

13. For the midpoint method (6.21), show that if $\lambda < 0$, then $0 < r_1 < 1$ and $r_2 < -1$.

    Solution: The key issues are that $\sqrt{\xi^2 + 1} > |\xi|$ and $\sqrt{\xi^2 + 1} \leq \sqrt{\xi^2 + 2|\xi| + 1} = |\xi| + 1$. Then

    $$r_1 > -|\xi| + |\xi| > 0$$

    and

    $$r_1 \leq -|\xi| + |\xi| + 1 = 1.$$

    Similarly,

    $$
    \begin{aligned}
    r_2 &= -|\xi| - \sqrt{|\xi|^2 + 1} < -|\xi| - \sqrt{|\xi|^2 - 2|\xi| + 1} = -|\xi| - \sqrt{(|\xi| - 1)^2} \\
    &= -|\xi| - \sqrt{(1 - |\xi|)^2} = -|\xi| - (1 - |\xi|) = -1.
    \end{aligned}
    $$

    We assumed at the end that h was small enough that $1 - |\xi| > 0$; this should have been an assumption in the problem statement.

14. Show that (6.44) is valid. Hint: First, show that $C_1 = 1 - C_2$. Next, use Taylor expansions to write

    $$\frac{\xi - e^{\xi}}{\sqrt{\xi^2 + 1}} = \frac{-1 - \frac{1}{2}\xi^2 + \mathcal{O}(\xi^3)}{1 + \frac{1}{2}\xi^2 + \mathcal{O}(\xi^4)}.$$

    Solution: Showing $C_1 = 1 - C_2$ is fairly routine. We then have that

    $$e^{\xi} = 1 + \xi + \frac{1}{2}\xi^2 + \mathcal{O}(\xi^3) \Rightarrow \xi - e^{\xi} = -1 - \frac{1}{2}\xi^2 + \mathcal{O}(\xi^3),$$

    and

    $$\sqrt{\xi^2 + 1} = 1 + \frac{1}{2}\xi^2 + \mathcal{O}(\xi^4).$$

    Both of these follow from routine Taylor series computations. Finally, we have

    $$
    \begin{aligned}
    \frac{-1 - \frac{1}{2}\xi^2 + \mathcal{O}(\xi^3)}{1 + \frac{1}{2}\xi^2 + \mathcal{O}(\xi^4)} &= \frac{-1 - \frac{1}{2}\xi^2 - \mathcal{O}(\xi^4) + \mathcal{O}(\xi^4) + \mathcal{O}(\xi^3)}{1 + \frac{1}{2}\xi^2 + \mathcal{O}(\xi^4)} \\
    &= -1 + \frac{\mathcal{O}(\xi^4) + \mathcal{O}(\xi^3)}{1 + \frac{1}{2}\xi^2 + \mathcal{O}(\xi^4)} \\
    &= -1 + \mathcal{O}(\xi^4) + \mathcal{O}(\xi^3) \\
    &= -1 + \mathcal{O}(\xi^3).
    \end{aligned}
    $$

15. Use the midpoint rule predictor-corrector method (6.35)–(6.36) to solve each of the IVP's given in Problem 4.

    Solution: A MATLAB script that does the job is given below.

```
function [tc,yc] = midpc(h,n,y0)
    tc = [0];
    yc = [y0];
    yn = y0;
    tn = 0;
    for k=1:n
```

```
yp = yn + 0.5*h*fmid(tn,yn);
y  = yn + h*fmid(tn+0.5*h,yp);
t = k*h;
tc = [tc t];
yc = [yc y];
yn = y;
tn = t;
end
tc = tc';
yc = yc'
exact = 0.5*(1 + 3*exp(-2*tc));
%
figure(1)
plot(tc,yc)
figure(2)
plot(tc,exact - yc)
```

16. Show that the residual for the midpoint rule predictor-corrector is given by

$$R = y(t + h) - y(t) - hf(t + h/2, y(t) + (h/2)f(t, y(t))).$$

Then use Taylor's theorem to show that $R = \mathcal{O}(h^3)$ and hence that the midpoint rule predictor-corrector is a second-order method.

Solution: We have

$$y(t + h) - y(t) = hy'(t) + \frac{1}{2}h^2 y''(t) + \mathcal{O}(h^3) = hf(t, y(t)) + \frac{1}{2}h^2 y''(t) + \mathcal{O}(h^3),$$

so

$$R = hf(t, y(t)) + \frac{1}{2}h^2 y''(t) + \mathcal{O}(h^3) - hf(t + h/2, y(t) + (h/2)f(t, y(t))).$$

For notational simplicity write $Y(t) = y(t) + (h/2)f(t, y(t))$ so that we have

$$R = hf(t, y(t)) + \frac{1}{2}h^2 y''(t) - hf(t + h/2, Y(t)) + \mathcal{O}(h^3).$$

The problem can be done from this point quite simply using Taylor's Theorem for two variable functions, but it can also be done using single variable calculus, carefully. We have

$$f(t + h/2, Y(t)) = f(t, Y(t)) + (h/2)f_t(t, Y(t)) + \mathcal{O}(h^2),$$

so R becomes

$$R = h(f(t, y(t)) - f(t, Y(t))) + (h^2/2)(y''(t) - f_t(t, Y(t))) + \mathcal{O}(h^3).$$

The first term simplifies as follows:

$$h(f(t, y(t)) - f(t, Y(t))) = hf_y(t, \eta_t)(y(t) - Y(t)) = -(h^2/2)f_y(t, \eta_t)f(t, y(t)),$$

where η_t is a value between $y(t)$ and $Y(t)$. Now, since

$$y'(t) = f(t, y(t))$$

it follows that

$$y''(t) = \frac{d}{dt}f(t, y(t)) = f_t(t, y(t)) + f_y(t, y(t))y'(t) = f_t(t, y(t)) + f_y(t, y(t))f(t, y(t)).$$

Therefore,

$$R = (h^2/2)(f_y(t, y(t))f(t, y(t)) - f_y(t, \eta_t)f(t, y(t)) + f_t(t, y(t)) - f_t(t, Y(t))) + \mathcal{O}(h^3).$$

Hence,

$$
\begin{aligned}
R &= (h^2/2)(f_y(t, y(t))f(t, y(t)) - f_y(t, \eta_t)f(t, y(t)) + f_t(t, y(t)) - f_t(t, Y(t))) + \mathcal{O}(\\
&= (h^2/2)f(t, y(t))[f_y(t, y(t)) - f_y(t, \eta_t)] + (h^2/2)[f_t(t, y(t)) - f_t(t, Y(t))] + \mathcal{O}(h^3 \\
&= (h^2/2)f(t, y(t))f_{yy}(t, \mu_t)[y(t) - \eta_t] + (h^2/2)f_{ty}(t, \nu_t)[y(t)) - Y(t)] + \mathcal{O}(h^3),
\end{aligned}
$$

where μ_t is a value between $y(t)$ and η_t and ν_t is a value between $y(t)$ and $Y(t)$.
Since

$$y(t) - Y(t) = -(h/2)f(t, y(t))$$

this quickly becomes

$$
\begin{aligned}
R &= (h^2/2)f(t, y(t))f_{yy}(t, \mu_t)[y(t) - \eta_t] - (h^3/4)f_{ty}()f(t, y(t)) + \mathcal{O}(h^3) \\
&= (h^2/2)f(t, y(t))f_{yy}(t, \mu_t)[y(t) - \eta_t] + \mathcal{O}(h^3).
\end{aligned}
$$

Finally, since η_t is between $y(t)$ and $Y(t)$ and $y(t) - Y(t) = \mathcal{O}(h)$, it follows that $y(t) - \eta_t = \mathcal{O}(h)$ which implies that $R = \mathcal{O}(h^3)$.

17. Assume that f is differentiable in y and that this derivative is bounded in absolute value for all t and y:

$$|f_y(t, y)| \le F.$$

Show that using fixed point iteration to solve for y_{n+1} in the trapezoid rule method will converge so long as h is sufficiently small. Hint: Recall Theorem 3.6.

Solution: The iteration function is

$$g(y) = Y + (h/2)(f(t, Y) + f(t + h, y)),$$

where we regard Y (which is equal to y_n) as a fixed value. From Theorem 3.6 we see that the convergence of the iteration depends on g' being less than one in absolute value. But

$$g'(y) = (h/2)\frac{\partial f}{\partial y}(t + h, y).$$

Under the given assumptions, we therefore have that

$$|g'(y)| \le hF/2 < 1$$

for h sufficiently small.

18. Derive the numerical method based on using Simpson's rule to approximate the integral in

$$y(t + h) = y(t - h) + \int_{t-h}^{t+h} f(s, y(s))ds.$$

What is the order of accuracy of this method? What is the truncation error? Is it implicit or explicit? Is it a single step or multistep method?

Solution: We get

$$y(t+h) = y(t-h) + \frac{h}{3}\left(f(t-h, y(t-h)) + 4f(t, y(t)) + f(t+h, y(t+h))\right)$$
$$- \frac{h^5}{2880} y^{(5)}(\xi_t),$$

which suggests the numerical method

$$y_{n+1} = y_{n-1} + \frac{h}{3}\left(f(t-h, y_{n-1}) + 4f(t, y_n) + f(t+h, y_{n+1})\right).$$

The method is an implicit multistep method, with order of accuracy 4; the truncation error is

$$\tau(t) = -\frac{h^4}{2880} y^{(5)}(\xi_t).$$

◁ • • ▷

6.5 SINGLE STEP METHODS—RUNGE-KUTTA

Exercises:

1. Use the method of Heun with $h = \frac{1}{4}$ to compute approximate values of $y(1)$ for each of the following initial value problems. Don't write a computer program, use a hand calculator (or calculator app) to produce an orderly table of (t_k, y_k) pairs.

   (a) $y' = y(1-y)$, $y(0) = \frac{1}{2}$;
   Solution:

   $$y_0 = 0.5, \quad y_1 = 0.5620117188, \quad y_2 = 0.6221229509, \quad y_3 = 0.6786683489,$$
   $$y_4 = 0.7303809855.$$

   (b) $y' = y(1 + e^{2t})$, $y(0) = 1$;
   Solution:

   $$y_0 = 1, \quad y_1 = 1.746635238, \quad y_2 = 3.674304010, \quad y_3 = 10.24009564,$$
   $$y_4 = 42.71053582.$$

   (c) $y' + 2y = 1$, $y(0) = 2$;
   Solution:

   $$y_0 = 2, \quad y_1 = 1.4375, \quad y_2 = 1.0859375, \quad y_3 = 0.8662109375,$$
   $$y_4 = 0.7288818360.$$

   (d) $ty' = y(\sin t)$, $y(0) = 2$.

2. Repeat the above, using fourth-order Runge-Kutta.

Solution:

(a)

$$y_0 = 2, \quad y_1 = 1.410156250, \quad y_2 = 1.052256267, \quad y_3 = .8350929950,$$
$$y_4 = .7033246558.$$

(b)

$$y_0 = 1, \quad y_1 = 1.775386270, \quad y_2 = 3.886047491, \quad y_3 = 11.97099553,$$
$$y_4 = 63.69087556.$$

(c)

$$y_0 = .5, \quad y_1 = 0.5621763730, \quad y_2 = 0.6224590537, \quad y_3 = 0.6791782275,$$
$$y_4 = 0.7310578607.$$

3. Write a computer program that solves each of the initial value problems in Problem 1, using the method of Heun, and $h = 1/16$.

Solution: Below is a MATLAB script that does this task.

```
function [tc, yc] = heun(y0,h,n)
    yc = [y0];
    tc = [0];
    y = y0;
    t = 0;
    for k=1:n
        k1 = h*fheun(t,y);
        k2 = h*fheun(t+2*h/3,y+(2/3)*k1);
        yn = y + 0.25*(k1 + 3*k2);
        y = yn;
        t = t + h;
        yc = [yc yn];
        tc = [tc t];
    end
    plot(tc,yc)
```

4. Write a computer program that solves each of the initial value problems in Problem 1, using fourth-order Runge-Kutta, and $h = 1/16$.

Solution: Below is a MATLAB script that does this task.

```
function [tc, yc] = rk4(y0,h,n)
    yc = [y0];
    tc = [0];
    y = y0;
    t = 0;
    for k=1:n
        k1 = h*frk(t,y);
        k2 = h*frk(t + 0.5*h,y + 0.5*k1);
```

```
        k3 = h*frk(t + 0.5*h,y + 0.5*k2);
        k4 = h*frk(t +     h,y +     k3);
        yn = y + (k1 + 2*k2 + 2*k3 + k4)/6;
        y = yn;
        t = t + h;
        yc = [yc yn];
        tc = [tc t];
    end
    plot(tc,yc)
```

5. For each initial value problem below, approximate the solution using the method of Heun with a sequence of decreasing grids $h^{-1} = 2, 4, 8, \ldots$. For those problems where an exact solution is given, compare the accuracy achieved over the interval $[0, 1]$ with the theoretical accuracy.

  (a) $y' + \sin y = 0, y(0) = 1$.

  (b) $y' + 4y = 1, y(0) = 1; y(t) = \frac{1}{4}(3e^{-4t} + 1)$.

  (c) $y' + y = \sin 4\pi t, y(0) = \frac{1}{2}$.

  (d) $y' = -y \ln y, y(0) = 3; y(t) = e^{(\ln 3)e^{-t}}$.

Solution: The script given above can be used to do this problem.

6. Repeat the above, using fourth-order Runge-Kutta as the numerical method.

Solution: The rk4 script given above can be used to do this problem.

7. Repeat Problem 8 from §6.4, except this time use fourth-order Runge Kutta to solve the differential equation, with $h = 1/8$.

Solution: The same general results are obtained as previously.

8. Repeat Problem 9 from §6.4, except this time use fourth-order Runge Kutta to solve the differential equation, with $h = 1/8$.

Solution: The same general results are obtained as previously.

9. Repeat Problem 11 from §6.4, except this time use fourth-order Runge Kutta to solve the differential equation, with $h = 1/8$.

Solution: The same general results are obtained as previously.

10. Repeat Problem 10 from §6.4, except this time use fourth-order Runge Kutta to solve the differential equation, with $h = 1/8$.

Solution: The same general results are obtained as previously.

◁ ● ● ● ▷

6.6 MULTISTEP METHODS

Exercises:

1. Use second-order Adams-Bashforth with $h = \frac{1}{4}$ to compute approximate values of $y(1)$ for each of the following initial value problems. Don't write a computer program, use a hand calculator (or calculator app) to produce an orderly table of (t_k, y_k) pairs. Use Euler's method to generate the needed starting value.

   (a) $y' + 2y = 1$, $y(0) = 2$;
   Solution:

   $$y_0 = 2, \quad y_1 = 1.25, \quad y_2 = 1.0625, \quad y_3 = 0.828125, \quad y_4 = 0.7226562500$$

   (b) $y' = y(1 + e^{2t})$, $y(0) = 1$;
   Solution:

   $$y_0 = 1, \quad y_1 = 1.5, \quad y_2 = 2.739905715, \quad y_3 = 6.063673588,$$
   $$y_4 = 17.25489584$$

   (c) $ty' = y(\sin t)$, $y(0) = 2$.
   (d) $y' = y(1 - y)$, $y(0) = \frac{1}{2}$;
   Solution:

   $$y_0 = .5, \quad y_1 = 0.5625, \quad y_2 = 0.6235351563, \quad y_3 = 0.6808005870,$$
   $$y_4 = 0.7329498843$$

2. Verify that the λ_k values and ρ_p are correct for second-order Adams-Bashforth.
   Solution: We have

   $$\lambda_0 = \int_{t_n}^{t_{n+1}} L_0(s)ds = \int_{t_n}^{t_{n+1}} \frac{s - t_{n-1}}{t_n - t_{n-1}}ds = \int_0^h \frac{u + h}{h}ds = \frac{3}{2}h$$

   and

   $$\lambda_1 = \int_{t_n}^{t_{n+1}} L_1(s)ds = \int_{t_n}^{t_{n+1}} \frac{s - t_n}{t_{n-1} - t_n}ds = \int_0^h \frac{u}{-h}ds = -\frac{1}{2}h$$

   and

   $$\rho_1 = \int_{t_n}^{t_{n+1}} \frac{(s - t_n)(s - t_{n-1})}{2}ds = \frac{1}{2}\int_0^h u(u + h)ds = \frac{5}{12}h^3$$

3. Write a computer program that solves each of the initial value problems in Problem 1, using second-order Adams-Bashforth and $h = 1/16$. Use Euler's method to generate the starting value.
   Solution: A MATLAB script that does this is given below.

```
function [tc, yc] = AB2(y0,h,n)
    yc = [y0];
```

```
      tc = [0];
      y = y0;
      t = 0;
%
      fy = fAB2(t,y);
      yn = y + h*fy;
      tn = h;
      tc = [tc tn];
      yc = [yc yn];
%
      for k=2:n
          fyn = fAB2(tn,yn);
          yp = yn + 0.5*h*(3*fyn - fy);
          y = yn;
          fy = fyn;
          yn = yp;
          tn = k*h;
          tc = [tc tn];
          yc = [yc yp];
      end
      plot(tc,yc)
```

4. Write a computer program that solves each of the initial value problems in Problem 1, using fourth-order Adams-Bashforth, with fourth-order Runge-Kutta to generate the starting values.

Solution: A MATLAB script which does this is given below.

```
function [t,y] = AB4(y0,h,n)
      t0 = 0;
      y = [y0];
      t = [0];
%
      k1 = fAB4(t0,y0);
      k2 = fAB4(t0+0.5*h,y0+0.5*h*k1);
      k3 = fAB4(t0+0.5*h,y0+0.5*h*k2);
      k4 = fAB4(t0+h,y0+h*k3);
      ya = y0 + (h/6)*(k1 + 2*k2 + 2*k3 + k4);
      ta = t0 + h;
      y = [y ya];
      t = [t ta];
%
      k1 = fAB4(ta,ya);
      k2 = fAB4(ta+0.5*h,ya+0.5*h*k1);
      k3 = fAB4(ta+0.5*h,ya+0.5*h*k2);
      k4 = fAB4(ta+h,ya+h*k3);
      yb = ya + (h/6)*(k1 + 2*k2 + 2*k3 + k4);
      tb = ta + h;
      y = [y yb];
      t = [t tb];
%
      k1 = fAB4(tb,yb);
      k2 = fAB4(tb+0.5*h,yb+0.5*h*k1);
```

```
        k3 = fAB4(tb+0.5*h,yb+0.5*h*k2);
        k4 = fAB4(tb+h,yb+h*k3);
        yc = yb + (h/6)*(k1 + 2*k2 + 2*k3 + k4);
        tc = tb + h;
        y = [y yc];
        t = [t tc];
  %
        for k=4:n
            yn = yc + (h/24)*(55*fAB4(tc,yc) - 59*fAB4(tb,yb) ...
                + 37*fAB4(ta,ya) - 9*fAB4(t0,y0));
            y0 = ya;
            ya = yb;
            yb = yc;
            yc = yn;
            t0 = ta;
            ta = tb;
            tb = tc;
            tn = tc + h;
            tc = tn;
            y = [y yn];
            t = [t tn];
        end
        figure(1)
        plot(t,y);
        ye = 1./(1+exp(-t));
        figure(2)
        plot(t,ye - y)
        ee = max(abs(ye - y))
```

5. For each initial value problem below, approximate the solution using the fourth-order Adams-Bashforth-Moulton predictor-corrector, with fourth-order Runge-Kutta to generate the starting values. Use a sequence of decreasing grids $h^{-1} = 2, 4, 8, \ldots$. For those problems where an exact solution is given, compare the accuracy achieved over the interval $[0, 1]$ with the theoretical accuracy.

   (a) $y' + \sin y = 0$, $y(0) = 1$.
   (b) $y' + 4y = 1$, $y(0) = 1$; $y(t) = \frac{1}{4}(3e^{-4t} + 1)$.
   (c) $y' + y = \sin 4\pi t$, $y(0) = \frac{1}{2}$.
   (d) $y' = -y \ln y$, $y(0) = 3$; $y(t) = e^{(\ln 3)e^{-t}}$.

   Solution: A MATLAB script that does this is given below. It is not especially efficient.

```
function [t,y] = ABAM4(y0,h,n)
    t0 = 0;
    y = [y0];
    t = [0];
%
    k1 = fABAM4(t0,y0);
    k2 = fABAM4(t0+0.5*h,y0+0.5*h*k1);
    k3 = fABAM4(t0+0.5*h,y0+0.5*h*k2);
```

```
k4 = fABAM4(t0+h,y0+h*k3);
ya = y0 + (h/6)*(k1 + 2*k2 + 2*k3 + k4);
ta = t0 + h;
y = [y ya];
t = [t ta];
%
k1 = fABAM4(ta,ya);
k2 = fABAM4(ta+0.5*h,ya+0.5*h*k1);
k3 = fABAM4(ta+0.5*h,ya+0.5*h*k2);
k4 = fABAM4(ta+h,ya+h*k3);
yb = ya + (h/6)*(k1 + 2*k2 + 2*k3 + k4);
tb = ta + h;
y = [y yb];
t = [t tb];
%
k1 = fABAM4(tb,yb);
k2 = fABAM4(tb+0.5*h,yb+0.5*h*k1);
k3 = fABAM4(tb+0.5*h,yb+0.5*h*k2);
k4 = fABAM4(tb+h,yb+h*k3);
yc = yb + (h/6)*(k1 + 2*k2 + 2*k3 + k4);
tc = tb + h;
y = [y yc];
t = [t tc];
%
for k=4:n
    yp = yc + (h/24)*(55*fABAM4(tc,yc) - 59*fABAM4(tb,yb) ...
            + 37*fABAM4(ta,ya) - 9*fABAM4(t0,y0));
    yn = yc + (h/24)*(9*fABAM4(tc+h,yp) + 19*fABAM4(tc,yc) ...
            - 5*fABAM4(tb,yb) + fABAM4(ta,ya));
    y0 = ya;
    ya = yb;
    yb = yc;
    yc = yn;
    t0 = ta;
    ta = tb;
    tb = tc;
    tn = tc + h;
    tc = tn;
    y = [y yn];
    t = [t tn];
end
figure(1)
plot(t,y,'k-');
ye = (16*exp(-t)*pi^2-8*pi*cos(4*pi*t)+8*exp(-t)*pi...
        +exp(-t)+2*sin(4*pi*t))/(32*pi^2+2);
figure(2)
plot(t,ye - y,'k-')
ee = max(abs(ye - y))
```

6. Derive the second order Adams-Bashforth method under the assumption that the grid is *not* uniform. Assume that $t_{n+1} - t_n = h$, and $t_n - t_{n-1} = \eta$, with $\eta = \theta h$. What is the truncation error in this instance?

Solution: $y_{n+1} = y_n + (h/2\theta)((1+2\theta)f(t_n, y_n) - f(t_{n-1}, y_{n-1}))$. The truncation error is

$$\rho(\theta) = h^3 \left(\frac{1}{6} + \frac{1}{4}\theta \right)$$

7. Repeat Problem 8 from §6.4, this time using second order Adams-Bashforth with $h = 1/16$. Use simple Euler to generate the starting value y_1. If you did the earlier problems of this type, compare your results now to what you got before.

   Solution: The same general results are obtained as previously.

8. Repeat Problem 9 from §6.4, this time using second order Adams-Bashforth with $h = 1/16$. Use simple Euler to generate the starting value y_1. If you did the earlier problems of this type, compare your results now to what you got before.

   Solution: The same general results are obtained as previously.

9. Repeat Problem 11 from §6.4, this time using second order Adams-Bashforth with $h = 1/16$. Use simple Euler to generate the starting value y_1. If you did the earlier problems of this type, compare your results now to what you got before.

   Solution: The same general results are obtained as previously.

10. Repeat Problem 10 from §6.4, this time using second order Adams-Bashforth with $h = 1/16$. Use simple Euler to generate the starting value y_1. If you did the earlier problems of this type, compare your results now to what you got before.

    Solution: The same general results are obtained as previously.

6.7 STABILITY ISSUES

Exercises:

1. Determine the residual and the truncation error for the method defined by $y_{n+1} = 4y_n - 3y_{n-1} - 2hf(t_{n-1}, y_{n-1})$. Try to use it to approximate solutions of the IVP $y' = -y \ln y$, $y(0) = 2$. Comment on your results.

   Solution: The truncation error for this method will indicate it is an $\mathcal{O}(h^2)$ method, yet the actual computation will be horribly inaccurate, because the method is unstable.

2. Show that the method defined by $y_{n+1} = 4y_n - 3y_{n-1} - 2hf(t_{n-1}, y_{n-1})$ is unstable.

   Solution: The stability polynomial is $\sigma(r) = r^2 - 4r + 3$, which has roots $r_0 = 1$ and $r_1 = 3$, so the method fails to be stable.

3. Consider the method defined by

   $$y_{n+1} = y_{n-1} + \frac{1}{3}h\left[f(t_{n-1}, y_{n-1}) + 4f(t_n, y_n) + f(t_{n+1}, y_{n+1})\right],$$

   which is known as Milne's method. Is this method stable or strongly stable?

Solution: The stability polynomial is $\sigma = r^2 - 1$, which has roots $r_0 = 1$ and $r_1 = -1$, so it is only weakly stable, like the midpoint method.

4. Show for backward Euler and the trapezoid rule methods, that the stability region is the entire left half-plane.

Solution: We do this only for the trapezoid rule method.

We have

$$y_{n+1} = y_n + (h/2)(\lambda y_n + \lambda y_{n+1})$$

so that

$$y_{n+1} = \left(\frac{1 + (h\lambda)/2}{1 - (h\lambda)/2}\right) y_n.$$

If $\Re(\lambda) < 0$, then $|y_{n+1}| = \gamma |y_n|$, where $\gamma < 1$, so $y_n \to 0$ as $n \to \infty$, hence the stability region is the entire left half-plane.

5. Show that all four members of the BDF family are stable.

Solution: The stability polynomials are

$$r = 1,$$

$$r^2 = \frac{4}{3}r - \frac{1}{3},$$

$$r^3 = \frac{18}{11}r^2 - \frac{9}{11} + \frac{2}{11},$$

and

$$r^4 = \frac{48}{25}r^3 - \frac{36}{25}r^2 + \frac{16}{25} - \frac{3}{25}.$$

Simple computations show that all four cases satisfy the root condition, and hence are stable.

6. Show that the stability region for the fourth-order BDF method contains the entire negative real axis.

Solution: The stability region is that part of the complex plane for which the method is relatively stable when applied to the ODE $y' = \lambda y$. Thus, we look at the recursion

$$y_{n+1} = \frac{48}{25}y_n - \frac{36}{25}y_{n-1} + \frac{16}{25}y_{n-2} - \frac{3}{25}y_{n-3} + \frac{12}{25}h\lambda y_{n+1}.$$

To say that the stability region for the method contains the entire negative real axis is equivalent to saying that the roots of the polynomial

$$r^4 = \frac{48}{25}r^3 - \frac{36}{25}r^2 + \frac{16}{25}r - \frac{3}{25} + \frac{12}{25}h\lambda r^4$$

are less than one in absolute value for all negative real λ. We thus look at the polynomial equation

$$(25 + 12\xi)r^4 - 48r^3 + 36r^2 - 16r + 3 = 0,$$

where $\xi = -h\lambda > 0$. For specific values of $\xi > 0$, we can compute the maximum absolute value of the roots of this polynomial. Since the maximum absolute value is always less than one, the negative real axis is within the stability region of the method.

◁ • • ▷

6.8   APPLICATION TO SYSTEMS OF EQUATIONS

Exercises:

1. Using a hand calculator (or calculator app) and $h = \frac{1}{4}$, compute approximate solutions to the initial value problem

$$y_1' = -4y_1 + y_2, \quad y_1(0) = 1;$$
$$y_2' = \quad y_1 - 4y_2, \quad y_2(0) = -1.$$

Compute out to $t = 1$, using the following methods:

(a) Euler's method;
   Solution: $y_{1,4} = 0.00390625$, $y_{2,4} = -0.00390625$

(b) RK4;

(c) AB2 with Euler's method for the starting values;
   Solution: $y_1(1) \approx 1.310058594$, $y_2(1) \approx -1.310058594$.

(d) Trapezoid rule predictor-corrector.
   Solution: $y_{1,4} = 0.07965183255$, $y_{2,4} = -0.07965183255$.

Organize your computations and results neatly to minimize mistakes.

2. Consider the second order equation

$$y'' + \sin y = 0, \quad y(0) = \frac{1}{8}\pi, y'(0) = 0;$$

write this as a first order system and compute approximate solutions using a sequence of grids and the following methods:

(a) Second-order Adams-Bashforth;

(b) Fourth-order Runge-Kutta;

(c) The method of Heun.

The solution will be periodic; try to determine, experimentally, the period of the motion.

Solution: As a system, the problem becomes

$$\frac{dy_1}{dt} = y_2, \quad \frac{dy_2}{dt} = -\sin y_1.$$

Using Heun's method and $h = 1/32$ we get that the period of the motion is about 6.3.

3. Show that backward Euler and the trapezoid rule methods are both A-stable.

Solution: We do this for backward Euler, only. We apply the method to the ODE $y' = \lambda y$, where $\Re(\lambda) < 0$ is assumed. This yields

$$y_{n+1} = y_n + h\lambda y_{n+1},$$

so that

$$y_{n+1} = \left(\frac{1}{1 - h\lambda} \right) y_n = \gamma y_n.$$

Since $\mathfrak{R}(\lambda) < 0$ it follows that $\gamma < 1$ so $y_n \to 0$, thus the method is A-stable. A similar argument works for the trapezoid rule method.

4. Show that the second-order BDF method is A-stable.

5. Apply the third-order BDF method, using the trapezoid method to obtain the starting values, to the stiff example in the text. Compare your results to the exact solution and those obtained in the text.

   Solution: At each step we have to solve the system

   $$\left(I - \frac{6}{11} hA \right) y_{n+1} = \frac{18}{11} y_n - \frac{9}{11} y_{n-1} - \frac{2}{11} y_{n-2},$$

   where

   $$y_n = \left(\begin{array}{c} y_{1,n} \\ y_{2,n} \end{array} \right)$$

   and

   $$A = \left[\begin{array}{cc} 198 & 199 \\ -398 & -399 \end{array} \right].$$

   Since we are using a third order method instead of a second order method, we expect better results and we get them. The maximum error for this method is substantially smaller than for the trapezoid rule.

6. Consider the following system of differential equations:

   $$\begin{array}{rcl} y_1'(t) &=& ay_1(t) - by_2(t)y_1(t), y_1(0) = y_{10}, \\ y_2'(t) &=& -cy_2(t) + dy_1(t)y_2(t), y_2(0) = y_{20}. \end{array}$$

   Solve this system for a sequence of mesh sizes, using the data $a = 4, b = 1, c = 2, d = 1$, with initial values $y_{10} = 3/2$ and $y_{20} = 4$. Use (a) Second-order Adams-Bashforth, with Euler's method providing the starting value; and, (b) Fourth-order Runge-Kutta. Plot the solutions in two ways: As a single plot showing y_1 and y_2 versus t; and as a *phase plot* showing y_1 versus y_2. Note: The solutions should be periodic. Try to determine, experimentally, the period. (Note: This problem is an example of a *predator-prey* model, in which y_1 represents the population of a prey species and y_2 represents the population of a predator which uses the prey as its only food source. See Braun [6] for an excellent discussion of the dynamics of such systems, as well as the history behind the problem.)

   Solution: Using AB2 with $h = 1/32$ you should get that the period of the solution is about 2.2.

7. Consider now a situation in which two species, denoted x_1 and x_2, *compete* for a common food supply. A standard model for this is

   $$\begin{array}{rcl} x_1'(t) &=& ax_1(t) - (bx_1(t) + cx_2(t))x_1(t) \\ x_2'(t) &=& Ax_2(t) - (Bx_1(t) + Cx_2(t))x_2(t) \end{array}$$

with initial conditions $x_1(0) = x_1^0$, $x_2(0) = x_2^0$. Using any of the methods in this chapter, solve this system for $x_1^0 = x_2^0 = 10,000$, using $a = 4, b = 0.0003, c = 0.004, A = 2, B = 0.0002, C = 0.0001$. Vary the parameters of the problem slightly and observe what happens to the solution.

8. An interesting and somewhat unusual application of systems of differential equations is to *combat modeling*. Here, we denote the force levels of the two sides at war by x_1 and x_2, and make various hypotheses about how one side's force level affects the losses suffered by the other side. For example, if we hypothesize that losses are proportional to the size of the opposing force, then (in the absence of reinforcements) we get the model

$$
\begin{aligned}
x_1' &= -ax_2, \\
x_2' &= -bx_1.
\end{aligned}
$$

If we hypothesize that losses are proportional to the *square* of the opposing force, we get the model

$$
\begin{aligned}
x_1' &= -\alpha x_2^2, \\
x_2' &= -\beta x_1^2.
\end{aligned}
$$

The constants α, β, a, b represent the military efficiency of one side's forces.

Using any of the methods of this section, consider both models with

$$
a = 1, \alpha = 1, x_1(0) = 120, x_2(0) = 40
$$

and observe how changing b and β affects the long-term trend of the solutions. In particular, can you find values of b and β such that the smaller force annihilates the larger one? Combat models such as this are known as *Lanchester models* after the British mathematician F. W. Lanchester who introduced them.

Solution: Using Euler's method with $\Delta t = 0.05$, the author found that, in the linear model, the smaller force needed to be have $b = 11$ to annihilate the larger force, while for the squared model it required $\beta = 37$ for the smaller force to win. (In the squared model the author used a much smaller time step, $\Delta t = 0.0005$, to avoid one side eliminating the other in a single step.)

9. Consider the following system of ODEs:

$$
\begin{aligned}
x' &= \sigma(y - x), \\
y' &= x(\rho - z) - y, \\
z' &= xy - \beta z.
\end{aligned}
$$

Use two different second order methods (your choice) to approximate solutions to this system, for $\sigma = 10$, $\beta = 8/3$, and $\rho = 28$. Take $x(0) = y(0) = z(0) = 1$ and compute out to $t = 5$. Plot each component.

10. Consider the second order equation:

$$
x'' - \mu(1 - x^2)x' + x = 0, \quad x(0) = 0.5, \quad x'(0) = 0.
$$

First, write this as a system. Then, as in the previous problem, choose two different second order methods and use each to solve the system for $\mu = 0.1$ and $\mu = 10$. Compute out to $t = 20$.

11. One simple but important application of systems of differential equations is to the spread of epidemics. Perhaps the simplest model is the so-called SIR model, which models an infectious disease (like measles) that imparts immunity to those who have had it and recovered. We divide the population into three categories:

- $S(t)$: These are the people who are susceptible, i.e., they are well at time t but might get sick at some future time;
- $I(t)$: These are the people who are infected, i.e., sick;
- $R(t)$: These are the people who have recovered and are therefore protected from getting the disease by being immune.

Under a number of simplifying assumptions[1] the SIR model involves the following system, sometimes known as the continuous Kermack-McKendrick model [22].

$$\frac{dS}{dt} = -rSI,$$

$$\frac{dI}{dt} = rSI - aI,$$

$$\frac{dR}{dt} = aI,$$

The parameters r and a measure the rate at which susceptible people become sick upon contact with infectives, and the rate at which infected people are removed (die or are cured).

(a) For the case $I_0 = 1$, $S_0 = 762$, $R_0 = 0$, $a = 0.44$, and $r = 2.18 \times 10^{-3}$, solve the system using the numerical method and step size of your choice, and plot the solutions. This data is taken from a study of an influenza epidemic at an English boys school; the time scale here is in days.

(b) Consider now the case $I_0 = 10$, $S_0 = 7.8 \times 10^6$, $R_0 = 0$, $a = 1/3$, and $r = 6.4 \times 10^{-8}$. These data are taken from the H3N2 flu epidemic that hit New York City in 1968–1969, and the time scale is in days. Again, choose the method of your choice and solve the system over a period of 140 days, and plot the solutions. Assume that the number of deaths was 0.5% of the infected population, and add a plot of the fatalities to your graph.

(c) In the SIR model, the parameter a can be interpreted as the reciprocal of the mean time of recovery; thus in (b), above, $a = 1/3$ means the disease usually lasts 3 days. Repeat the computation from (b), but this time assume a disease that typically lasts 14 days, and whose fatality rate is 3.4%. Is there any significant change from the previous case?

---

[1] See J.D. Murray, *Mathematical Biology*, Springer-Verlag, Berlin, 1989, pp. 611ff.

6.9 ADAPTIVE SOLVERS

Exercises:

1. Use Algorithm 6.4 to solve the following IVP using the local error tolerance of $\epsilon = 0.001$:

$$y' = y^2, \quad y(0) = 1.$$

Do this as a hand calculation. Compute 4 or 5 steps. Compare the computed solution to the exact solution $y = (1-t)^{-1}$.

Solution: The algorithm settles on $h_1 = h_2 = h_3 = h_4 = h_5 = 0.0078125$ and produces the approximate values

$$y_1 = 1.00785339019122, \quad y_2 = 1.01583078044887, \quad y_3 = 1.02393512571904,$$

$$y_4 = 1.03216947546184, \quad y_5 = 1.04053697745659.$$

2. Apply one of the MATLAB routines to each of the following ODEs:

   (a) $y' + 4y = 1, y(0) = 1$;
   (b) $y' + \sin y = 0, y(0) = 1$;
   (c) $y' + y = \sin 4\pi t, y(0) = \frac{1}{2}$;
   (d) $y' = \frac{y-t}{y+t}, y(0) = 4$.

   Use both $\tau = 10^{-3}$ and $\tau = 10^{-6}$, and compute out to $t = 10$. Plot your solutions. You may need to adjust `MaxStep`.

   Solution: For (b), using an old version of `ode45`, the author got the graph in Fig. 6.1.

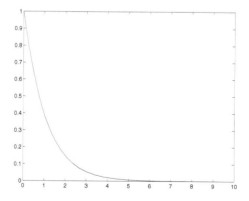

Figure 6.1   Solution to Problem 2b.

3. Repeat the previous problem, this time using $(\tau_r, \tau_a) = (10^{-6}, 10^{-9})$, and compare your results from the two cases.

4. Write up a few paragraphs outlining the advantages/disadvantages of the two criteria used here for governing the acceptance of computed values.

5. The MATLAB ODE routines can be applied to systems. Use each of ode23, ode45, and ode113 to solve the system in Problem 9 of §6.8, using $\tau = 10^{-3}$ and also $\tau = 10^{-6}$.

Solution: Using an old version of ode23, and plotting each component as a co-ordinate in 3-space, we get the graph in Fig. 6.2 (computing out to $T_{\max} = 65$). This system is known as the "Lorenz attractor," and is one of the first examples of a so-called chaotic dynamical system.

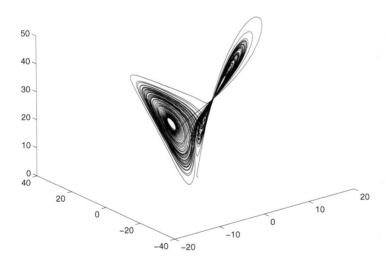

Figure 6.2   Solution to Problem 3.

6. Apply ode45 to the system in Problem 10 of §6.8, using $\mu = 10$. Compute out to $t = 60$, and plot both solution components. Comment.

Solution: Using the version of ode45 from the 1994 MATLAB, the author got the plot in Figure 6.3. This equation is known as the van der Pol oscillator; it is an example of a *nonlinear oscillator*. Note to instructors: Advise the students to plot the solution components on separate axes; otherwise the large values of the derivative will distort the scales.

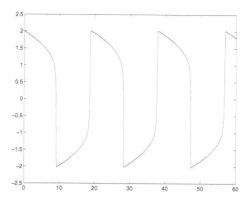

Figure 6.3   Solution curve for Problem 10.

7. The three routines we used in this section are not designed for stiff ODEs, but that does not mean we can't try it! Apply `ode23` to the system

$$
\begin{aligned}
y_1' &= 198y_1 + 199y_2, \quad y_1(0) = 1; \\
y_2' &= -398y_1 - 399y_2, \quad y_2(0) = -1;
\end{aligned}
$$

using $\tau = 10^{-6}$. Plot the solution components over the interval $(0, 5)$.

◁ • • ▷

6.10   BOUNDARY VALUE PROBLEMS

Exercises:

1. Use a mesh size of $h = \frac{1}{4}$ ($n = 4$) to solve the boundary value problem

$$
-u'' + u = \sin x, 0 < x < 1;
$$
$$
u(0) = u(1) = 0.
$$

Don't bother to write a computer program; do this as a hand calculation.

Solution: The system is

$$
\begin{bmatrix}
33 & -16 & 0 \\
-16 & 33 & -16 \\
0 & -16 & 33
\end{bmatrix}
\begin{bmatrix}
u_1 \\
u_2 \\
u_3
\end{bmatrix}
=
\begin{bmatrix}
\sqrt{2}/2 \\
1/2 \\
\sqrt{2}/2
\end{bmatrix},
$$

which has solution $u_1 = 0.05430593375937$, $u_2 = 0.06781181455454$, $u_3 = 0.05430593375937$.

2. Repeat the above for the BVP

$$
-u'' + (2 - x)u = x, 0 < x < 1;
$$
$$
u(0) = u(1) = 0.
$$

Don't bother to write a computer program; do this as a hand calculation.

Solution: $u(1/4) \approx 0.0329$, $u(1/2) \approx 0.0538$, $u(3/4) \approx 0.0484$.

3. Repeat the above for the BVP

$$-u'' + u = e^{-x}, 0 < x < 1;$$
$$u(0) = u(1) = 0.$$

Solution:

$$u_1 = 0.0572, \quad u_2 = 0.0694, \quad u_3 = 0.0480.$$

4. Consider the linear boundary value problem

$$-u'' + (10 \cos 2x)u = 1,$$
$$u(0) = u(\pi) = 0.$$

Solve this using finite difference methods and a decreasing sequence of grids, starting with $h = \pi/16, \pi/32, \ldots$. Do the approximate solutions appear to be converging to a solution?

Solution: Using $h = \pi/8$, we get

$$u = (0.0601, 0.0316, -0.1511, -0.3233, -0.1511, 0.0316, 0.0601)^T$$

as the solution vector; the approximate solutions do converge as $h \to 0$.

5. Develop the $\mathcal{O}(h^2)$ discrete tridiagonal system for the following BVP:

$$-u'' + u = f(x), \quad 0 < x < 1,$$
$$u(0) = 0, \quad u(1) + u'(1) = 1,$$

which has exact solution $u(x) = 1 - e^{-x}$ for $f(x) = 1$ (verify this.) Solve this problem using your discretization, and confirm that the code is working—that it delivers the expected accuracy—over a sequence of grids.

6. Consider the nonlinear boundary value problem

$$-u'' + e^u = 1,$$
$$u(0) = 0, \ u(1) = 1.$$

Use shooting methods combined with the trapezoid rule predictor-corrector to construct solutions to this equation, then use a fourth-order Runge-Kutta scheme and compare the two approximate solutions. Are they nearly the same?

Solution: To do this problem we first have to cast it into a first-order system:

$$\begin{pmatrix} w_1(t) \\ w_2(t) \end{pmatrix}' = \begin{pmatrix} w_2(t) \\ e^{w_1(t)} - 1 \end{pmatrix}.$$

We then solve this system using the initial values $w_1(0) = 0$, $w_2(t) = p$, and find the value of p such that $w_1(1) = 0$. Using the trapezoid rule predictor-corrector we get that $p = 0.8172$, and a similar value is obtained with the fourth-order Runge-Kutta method.

7. Repeat Problem 4 using Numerov's scheme.

8. Bogucz and Walker [4] developed an $\mathcal{O}(h^4)$ scheme for linear problems. If we restrict ourselves to problems of the form

$$-u'' + bu' + cu = f(x),$$
$$u(0) = g_0, \; u(1) = g_1,$$

for constant b and c, then the tridiagonal discretization has the form

$$A_k U_{k-1} + D_k U_k + B_k U_{k+1} = R_k,$$

for

$$D_k = 2 + \frac{1}{6} h^2 (5c + b^2);$$

$$A_k = -1 - \frac{1}{2} hb + \frac{1}{12} h^2 (c - b^2) + \frac{1}{24} h^3 g_0;$$

$$B_k = -1 + \frac{1}{2} hb + \frac{1}{12} h^2 (c - b^2) - \frac{1}{24} h^3 g_1;$$

$$R_k = h^2 f(x_k) + \frac{1}{12} h^4 (f''(x_k) + bf'(x_k));$$

where the boundary value terms are only applied to the appropriate equations near the boundary.

Write a script to solve Examples 6.31 and 6.32 using this scheme, and compare your results to those given in the text using the Numerov and Chawla methods.

Solution: A MATLAB function that sets up and solves the linear system is given below (the author wrote this assuming that the right-side function was a constant, but it is fairly easy to generalize beyond that):

```
function u = bvp4DBW(g0,g1,b,c,f,N)
%
% Solves 2-pt BVP
%
% -u" + bu' + cu = f
%
%
% Can be modified to handle first derivative term
%
% on (0,1), with n grid points (ASSUMES CONSTANT RHS!!!!))
%
% Bogusz-Walker scheme
%
h = 1/N;
h2 = h*h;
f6 = 5/6;
lambda = h2/12;
h24 = h^3/24;
sixth = 1/6;
bc = b*c;
vd = ones(N-1,1)';
```

```
vo = ones(N-2,1)';
D = 2 + sixth*h2*(5*c + b^2);
L = -1 - 0.5*h*b + lambda*(-b^2 + c) + h24*bc;
U = -1 + 0.5*h*b + lambda*(-b^2 + c) - h24*bc;
%
dd = D*vd;
l = L*[0 vo];
u = U*[vo 0];
%
ff = h2*f*ones(N-1,1)';
ff(1) = ff(1) - L*g0;
ff(N-1) = ff(N-1) - U*g1;
[z] = trisol(l,dd,u,ff);
    u = z;
end
```

Note that you still need to write a main program to use this, and you need a function trisol that solves tridiagonal systems using the material from §2.6 in the text. Assuming we have those, then we get the results in Table 6.2 for Example 6.31 and Table 6.3 for Example 6.32. Note that the error ratios are all about 16 until the error begins to saturate.

Table 6.2   Results for Problem 8, first part.

n	Error	Ratio
4	$1.6634629878 \times 10^{-6}$	NA
8	$1.0415676152 \times 10^{-7}$	15.97076333
16	$6.5128122062 \times 10^{-9}$	15.99259402
32	$4.0709610682 \times 10^{-10}$	15.99821786
64	$2.5433849471 \times 10^{-11}$	16.00607518
128	$1.5543955012 \times 10^{-12}$	16.36253415
256	$2.4910629115 \times 10^{-14}$	62.39888579
512	$6.9345917897 \times 10^{-13}$	0.03592227
1,024	$2.4131391330 \times 10^{-12}$	0.28736809
2,048	$7.8557160776 \times 10^{-12}$	0.30718258
4,096	$4.3876235978 \times 10^{-11}$	0.17904262

9. Write a program to use the finite element method to solve the BVP

$$-u'' + u = e^{-x}, 0 < x < 1;$$
$$u(0) = u(1) = 0;$$

using a sequence of grids, $h^{-1} = 4, 8, 16, \ldots, 1024$.

Solution: Using $n = 256$, the author got the plot in Figure 6.4; the values for $n = 16$ are marked by the circles.

10. Repeat the above, using the different boundary conditions $u(0) = 0, \quad u(1) = 1$.

11. No exact solution was provided for either of the previous two exercises (although anyone having completed a sophomore ODE course—or with access to a symbolic

Table 6.3 Results for Problem 8, second part.

n	Error	Ratio
4	$1.5100476030 \times 10^{-6}$	NA
8	$1.0116292228 \times 10^{-7}$	14.92688792
16	$6.3605568729 \times 10^{-9}$	15.90472726
32	$3.9981860878 \times 10^{-10}$	15.90860639
64	$2.4999835535 \times 10^{-11}$	15.99284956
128	$1.5290269051 \times 10^{-12}$	16.35016065
256	$9.2106877680 \times 10^{-14}$	16.60057255
512	$4.7595261066 \times 10^{-13}$	0.19352111
1,024	$4.5434489504 \times 10^{-13}$	1.04755796
2,048	$8.9535739933 \times 10^{-12}$	0.05074453
4,096	$1.8251788969 \times 10^{-12}$	4.90558707

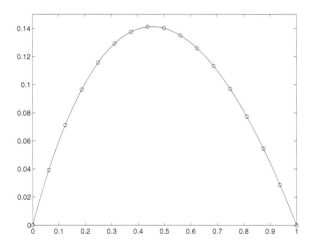

Figure 6.4 Solution plot for Problem 9.

algebra program like Maple or Mathematica—ought to be able to produce a solution). Write an essay addressing the following question: On what basis are you confidant that your codes are producing the correct solution?

Solution: If the sequence of solutions are not varying wildly from each other, if the results appear to be converging to a single solution, then we are justified in thinking—but by no means are we certain—that our code is correct. The fact that our solutions for $n = 16$ and $n = 256$ match as well as they do in Problem 9 gives us confidence that we are getting good results.

◁ ● ● ▷

CHAPTER 7

NUMERICAL METHODS FOR THE SOLUTION OF SYSTEMS OF EQUATIONS

7.1 LINEAR ALGEBRA REVIEW

Exercises:

1. Assume Theorem 7.1 and use it to prove Cor. 7.1.

   Solution: If A is singular, then Part 2 of the theorem says that the columns of A are a dependent set. Therefore, there exists a set of constants $c_1, c_2, \ldots, c_n$, not all zero, such that

   $$0 = c_1 a_1 + c_2 a_2 + \cdots + c_n a_n,$$

   where the a_i are the columns of A written as vectors. But this is equivalent to saying $Ac = 0$, where $c \in \mathbb{R}^n$ has components c_k. Therefore there exists at least one non-zero vector c such that $Ac = 0$. Now, let β be an arbitrary real number; then $A(\beta c) = \beta Ac = 0$, so there must be infinitely many non-zero vectors x such that $Ax = 0$.

2. Use Theorem 7.1 to prove that a triangular matrix is nonsingular if and only if the diagonal elements are all non-zero.

   Solution: Let T be the triangular matrix. Then $\det T = \prod t_{ii}$, so $\det T$ is non-zero if and only if all the diagonal elements are non-zero.

Solutions Manual to Accompany An Introduction to Numerical Methods and Analysis, Third Edition.
James F. Epperson.
© 2021 John Wiley & Sons, Inc. Published 2021 by John Wiley & Sons, Inc.

3. Suppose that we can write $A \in \mathbb{R}^{n \times n}$ as the product of two triangular matrices $L \in \mathbb{R}^{n \times n}$ and $U \in \mathbb{R}^{n \times n}$ where the diagonal elements of L and U are all non-zero. Prove that A is nonsingular.

Solution: Since L and U are triangular with non-zero diagonal elements, they are each non-singular. Therefore

$$Ax = b \Leftrightarrow LUx = b \Leftrightarrow Ux = L^{-1}b \Leftrightarrow x = U^{-1}L^{-1}b;$$

thus the system has a unique solution for any b, thus A is non-singular.

7.2   LINEAR SYSTEMS AND GAUSSIAN ELIMINATION

Exercises:

For the sake of simplicity here, we will define at the outset several families of matrices, parameterized by their dimension. These will be referred to in several of the exercises throughout the chapter.

$$H_n = [h_{ij}], \quad h_{ij} = \frac{1}{i + j - 1}.$$

$$K_n = [k_{ij}], \quad k_{i,j} = \begin{cases} 2, & i = j; \\ -1, & |i - j| = 1; \\ 0, & \text{otherwise.} \end{cases}$$

$$T_n = [t_{ij}], \quad t_{i,j} = \begin{cases} 4, & i = j; \\ 1, & |i - j| = 1; \\ 0, & \text{otherwise.} \end{cases}$$

$$A_n = [a_{ij}], \quad a_{i,j} = \begin{cases} 1, & i = j; \\ 4, & i - j = 1; \\ -4, & i - j = -1; \\ 0, & \text{otherwise.} \end{cases}$$

Even if we do not know the solution to a linear system, we can check the accuracy of a computed solution x_c by means of the residual $r = b - Ax_c$. If x_c is the exact solution, then each component of r will be zero; in floating point arithmetic, there might be a small amount of rounding error, unless the matrix is "nearly singular," a concept that we will discuss in detail in §7.5.

1. Write a naive Gaussian elimination code and use it to solve the system of equations $Ax = b$ where

$$A = \begin{bmatrix} 14 & 14 & -9 & 3 & -5 \\ 14 & 52 & -15 & 2 & -32 \\ -9 & -15 & 36 & -5 & 16 \\ 3 & 2 & -5 & 47 & 49 \\ -5 & -32 & 16 & 49 & 79 \end{bmatrix}$$

and $b = [-15, -100, 106, 329, 463]^T$. The correct answer is $x = [0, 1, 2, 3, 4]^T$.

Solution: A MATLAB script which does this is given below.

```
function x = naive(a,b)
    n = length(b);
    x = b;
    for i=1:(n-1)
        for j=(i+1):n
            m = a(j,i)/a(i,i);
            for k=(i+1):n
                a(j,k) = a(j,k) - m*a(i,k);
            end
            b(j) = b(j) - m*b(i);
        end
    end
%
    x(n) = b(n)/a(n,n);
    for i=(n-1):(-1):1
        s = 0;
        for j=(i+1):n
            s = s + a(i,j)*x(j);
        end
        x(i) = (b(i) - s)/a(i,i);
    end
```

2. Write a naive Gaussian elimination code and use it to solve the system of equations

$$T_5 x = b$$

where $b = [1, 6, 12, 18, 19]^T$. The correct answer is $x = [0, 1, 2, 3, 4]^T$.

Solution: The previous script will work.

3. Write a naive Gaussian elimination code and use it to solve the system of equations

$$H_5 x = b$$

where $b = [5.0, \ 3.550, \ 2.81428571428571, \ 2.34642857142857, \ 2.01746031746032]^T$. The correct answer is $x = [1, 2, 3, 4, 5]^T$.

Solution: The same script works, of course.

4. Repeat the above problem, except now use $b_1 = 5.0001$. How much does the answer change?

Solution: Now we get

$$x = [1.00249999999970, \ 1.96999999999926, \ 3.10500000002115,$$
$$3.85999999994903, \ 5.06300000003183]^T.$$

This is more change in the answer than we perhaps would have expected, given the small change in b. This is discussed in §7.5.

5. Write your own naive Gaussian elimination code, based on the material in this chapter, and test it on the indicated families, over the range of $4 \leq n \leq 20$. Take b to be the vector of appropriate size, each of whose entries is 1.

   (a) H_n;

   (b) K_n;

   (c) T_n.

   For each value of n, compute the value of $\max_{1 \leq i \leq n} |r_i|$, where $r = b - Ax$.

   Solution: For the second and third cases, the `naive` script returns a residual that is identically zero for all components. For the first case—known as the Hilbert matrix—the maximum element in the residual is still small, but non-zero. The author got $2.288243194925598 \times 10^{-8}$ as the largest element in the residual for $n = 20$.

6. Modify the Gaussian elimination algorithm to handle more than a single right hand side. Test it on a 5×5 example of your own design, using at least 3 right-hand side vectors.

   Solution: A MATLAB script which does this is given below. It is a modest change from the previous one.

```
function x = naive(a,b)
    [n,p] = size(b);
    x = b;
    for i=1:(n-1)
        for j=(i+1):n
            m = a(j,i)/a(i,i);
            for k=(i+1):n
                a(j,k) = a(j,k) - m*a(i,k);
            end
            for k = 1:p
                b(j,k) = b(j,k) - m*b(i,k);
            end
        end
    end
%
    for ii=1:p
    x(n,ii) = b(n,ii)/a(n,n);
    for i=(n-1):(-1):1
        s = 0;
        for j=(i+1):n
            s = s + a(i,j)*x(j,ii);
        end
        x(i,ii) = (b(i,ii) - s)/a(i,i);
    end
    end
```

7. Use the naive Gaussian elimination algorithm to solve (by hand) the following system. You should get the same results as in (7.1).

$$\begin{bmatrix} \epsilon & 1 \\ 1 & 1 \end{bmatrix} \begin{bmatrix} x_1 \\ x_2 \end{bmatrix} = \begin{bmatrix} 1 \\ 2 \end{bmatrix}.$$

Solution: We have

$$\left[\begin{array}{cc|c} \epsilon & 1 & 1 \\ 1 & 1 & 2 \end{array}\right] \sim \left[\begin{array}{cc|c} \epsilon & 1 & 1 \\ 0 & 1-(1/\epsilon) & 2-(1/\epsilon) \end{array}\right]$$

so that

$$x_2 = \frac{2-(1/\epsilon)}{1-(1/\epsilon)} = \frac{2\epsilon - 1}{\epsilon - 1} = \frac{1-2\epsilon}{1-\epsilon} = 1 - \frac{\epsilon}{1-\epsilon}$$

and

$$x_1 = \frac{1}{\epsilon}(1-x_2) = \frac{1}{\epsilon}\left(1-1+\frac{\epsilon}{1-\epsilon}\right) = \frac{1}{1-\epsilon} = 1 + \frac{\epsilon}{1-\epsilon}.$$

8. Write a Gaussian elimination code that does partial pivoting and use it to solve the system of equations $Ax = b$ where

$$A = \begin{bmatrix} 9 & 3 & 2 & 0 & 7 \\ 7 & 6 & 9 & 6 & 4 \\ 2 & 7 & 7 & 8 & 2 \\ 0 & 9 & 7 & 2 & 2 \\ 7 & 3 & 6 & 4 & 3 \end{bmatrix}$$

and $b = [35, 58, 53, 37, 39]^T$. The correct answer is $x = [0, 1, 2, 3, 4]^T$.

Solution: A MATLAB script that does the assigned task is given below.

```
function x = naivep(a,b)
    n = length(b);
    x = b;
    for i=1:(n-1)
        am = abs(a(i,i));
        p = i;
        for j=(i+1):n
            if abs(a(j,i)) > am
                am = abs(a(j,i));
                p = j;
            end
        end
        if p > i
            for k=i:n
                hold = a(i,k);
                a(i,k) = a(p,k);
                a(p,k) = hold;
            end
            hold = b(i);
            b(i) = b(p);
            b(p) = hold;
        end
        for j=(i+1):n
            m = a(j,i)/a(i,i);
            for k=(i+1):n
                a(j,k) = a(j,k) - m*a(i,k);
            end
            b(j) = b(j) - m*b(i);
```

```
        end
    end
%
    x(n) = b(n)/a(n,n);
    for i=(n-1):(-1):1
        s = 0;
        for j=(i+1):n
            s = s + a(i,j)*x(j);
        end
        x(i) = (b(i) - s)/a(i,i);
    end
```

9. Write a naive Gaussian elimination code and use it to solve the system of equations $Ax = b$ where

$$A = \begin{bmatrix} 1 & 1/2 & 1/3 \\ 1/2 & 1/3 & 1/4 \\ 1/3 & 1/4 & 1/5 \end{bmatrix}$$

and $b = [7/6, 5/6, 13/20]^T$. The correct answer is $x = [0, 1, 2]^T$.

Solution: The previous script will work.

10. Use the naive Gaussian elimination algorithm to solve (by hand) the following system, using only 3 digit decimal arithmetic. Repeat, using Gaussian elimination with partial pivoting. Comment on your results.

$$\begin{bmatrix} 0.0001 & 1 \\ 1 & 1 \end{bmatrix} \begin{bmatrix} x_1 \\ x_2 \end{bmatrix} \begin{bmatrix} 1 \\ 2 \end{bmatrix}.$$

Solution: This has to be done carefully, and it is important to remember that the restriction as to 3 digit arithmetic applies to *all* the computations. So we can't just apply the formulas from the previous problem.

If we don't pivot, then the elimination goes like this (assuming the machine chops):

$$\begin{bmatrix} 0.0001 & 1 & | & 1 \\ 1 & 1 & | & 2 \end{bmatrix} \sim \begin{bmatrix} 0.0001 & 1 & | & 1 \\ 0 & -9990 & | & -9990 \end{bmatrix}.$$

So, $x_2 = 1$ and $x_1 = 0$, not a very accurate result.

If we do pivot, then the elimination goes like this:

$$\begin{bmatrix} 0.0001 & 1 & | & 1 \\ 1 & 1 & | & 2 \end{bmatrix} \sim \begin{bmatrix} 1 & 1 & | & 2 \\ 0.0001 & 1 & | & 1 \end{bmatrix} \sim \begin{bmatrix} 1 & 1 & | & 2 \\ 0 & 0.999 & | & 0.999 \end{bmatrix}.$$

So, $x_2 = 1$ (again) but $x_1 = 2 - x_2 = 1$, which is a much more accurate result.

11. Write a code to do Gaussian elimination with partial pivoting, and apply it to the system $A_5 x = b$, where $b = [-4, -7, -6, -5, 16]^T$ and the solution is $x = [0, 1, 2, 3, 4]^T$.

Solution: The `naivep` script will work.

12. Use MATLAB's `rand` function to generate A, a random 10×10 matrix, and a random vector $b \in \mathbb{R}^{10}$; solve the system $Ax = b$ two different ways: (1) By using your

own code. (2) By using MATLAB's backslash command: x = A\b. Obviously, you should get the same results both times.

13. Repeat Problem 12, this time using a 20×20 random matrix and appropriate random right-hand side.

◁ ● ● ▷

7.3 OPERATION COUNTS

Exercises:

1. Determine the operation count for the tridiagonal solution algorithm of §2.6.

   Solution: About $C_1 = 4n$ for the elimination step, and $C_2 = 2n$ for the solution phase, for a total cost of $C = 6n$.

2. What is the operation count for computing the dot product $x \cdot y$ of two vectors?

   Solution: $C = n$.

3. What is the operation count for computing the matrix–vector product Ax?

4. Repeat Problem 5, assuming that A is tridiagonal.

5. What is the operation count for a matrix-matrix product, AB?

6. Repeat the above, assuming that A is tridiagonal.

   Solution: This time we have

   $$AB = (D_1 + D_2 + D_3)B,$$

   so the cost is roughly $C = 3n^2m + \mathcal{O}(1)$, assuming n is the length of the longest diagonal, and that B is $n \times m$.

7. What is the operation count for the *outer product* xy^T?

   Solution: $C = n^2$.

8. Determine the operation count for the backward solution algorithm.

   Solution:

   $$C = \sum_{i=1}^{n-1} \left(\sum_{j=i+1}^{n} 1 \right) = \frac{1}{2}n^2 + \mathcal{O}(n).$$

9. Repeat the above for the Gaussian elimination code you wrote in the previous section.

10. Repeat Problem 8 for the tridiagonal solver you wrote back in §2.6.

11. Assume that you are working on a computer that does one operation every 10^{-9} seconds. How long, roughly, would it take such a computer to solve a linear system for $n = 100,000$, using the cost estimates derived in this section for Gaussian

elimination? What is the time estimate if the computer only does one operation every 10^{-6} seconds?

Solution: About 92.6 hours for the fast computer, and $92,592.6$ hours (over 10 *years*) for the slow one.

◁ • • • ▷

7.4   THE *LU* FACTORIZATION

Exercises:

1. Do, by hand, an *LU* factorization of the matrix

$$A = \begin{bmatrix} 2 & 1 \\ 1 & 2 \end{bmatrix}$$

and use it to solve the system $Ax = b$, where $b = (1,2)^T$. The exact solution is $x = (0,1)^T$. Verify that $LU = A$.

Solution: We get

$$A = \begin{bmatrix} 2 & 1 \\ 1 & 2 \end{bmatrix} = \begin{bmatrix} 1 & 0 \\ 1/2 & 1 \end{bmatrix} \begin{bmatrix} 2 & 1 \\ 0 & 3/2 \end{bmatrix}$$

so

$$y = \begin{bmatrix} 1 \\ 3/2 \end{bmatrix}$$

and

$$x = \begin{bmatrix} 0 \\ 1 \end{bmatrix}$$

as expected.

2. Repeat Problem 1 for

$$A = \begin{bmatrix} 4 & 1 \\ 1 & 5 \end{bmatrix}$$

and $b = (2,10)^T$; here the exact solution is $x = (0,2)^T$.

Solution: We get

$$A = \begin{bmatrix} 4 & 1 \\ 1 & 5 \end{bmatrix} = \begin{bmatrix} 1 & 0 \\ 1/4 & 1 \end{bmatrix} \begin{bmatrix} 4 & 1 \\ 0 & 15/4 \end{bmatrix}.$$

Then $y = (2,19/2)^T$ and $x = (0,2)^T$, as expected.

3. Write an *LU* factorization code and use it to solve the system of equations $Ax = b$ where

$$A = \begin{bmatrix} 14 & 14 & -9 & 3 & -5 \\ 14 & 52 & -15 & 2 & -32 \\ -9 & -15 & 36 & -5 & 16 \\ 3 & 2 & -5 & 47 & 49 \\ -5 & -32 & 16 & 49 & 79 \end{bmatrix}$$

and $b = [-15, -100, 106, 329, 463]^T$. The correct answer is $x = [0, 1, 2, 3, 4]^T$.

Solution: It is an easy modification to the `naive` script to do this. You get

$$L = \begin{bmatrix} 1.0000 & 0 & 0 & 0 & 0 \\ 1.0000 & 1.0000 & 0 & 0 & 0 \\ -0.6429 & -0.1579 & 1.0000 & 0 & 0 \\ 0.2143 & -0.0263 & -0.1103 & 0.9140 & 1.0000 \\ -0.3571 & -0.7105 & 0.2912 & 1.0000 & 0 \end{bmatrix}$$

and

$$U = \begin{bmatrix} 14.0000 & 14.0000 & -9.0000 & 3.0000 & -5.0000 \\ 0 & 38.0000 & -6.0000 & -1.0000 & -27.0000 \\ 0 & 0 & 29.2669 & -3.2293 & 8.5226 \\ 0 & 0 & 0 & 50.3013 & 55.5483 \\ 0 & 0 & 0 & 0 & -0.4689 \end{bmatrix}.$$

4. Write an *LU* factorization code and use it to solve the system of equations

$$T_5 x = b,$$

where $b = [1, 6, 12, 18, 19]^T$. The correct answer is $x = [0, 1, 2, 3, 4]^T$.

Solution: You get

$$L = \begin{bmatrix} 1.0000 & 0 & 0 & 0 & 0 \\ 0.2500 & 1.0000 & 0 & 0 & 0 \\ 0 & 0.2667 & 1.0000 & 0 & 0 \\ 0 & 0 & 0.2679 & 1.0000 & 0 \\ 0 & 0 & 0 & 0.2679 & 1.0000 \end{bmatrix}$$

and

$$U = \begin{bmatrix} 4.0000 & 1.0000 & 0 & 0 & 0 \\ 0 & 3.7500 & 1.0000 & 0 & 0 \\ 0 & 0 & 3.7333 & 1.0000 & 0 \\ 0 & 0 & 0 & 3.7321 & 1.0000 \\ 0 & 0 & 0 & 0 & 3.7321 \end{bmatrix}.$$

5. Write an *LU* factorization code and use it to solve the system of equations

$$H_5 x = b,$$

where $b = [5.0, 3.550, 2.81428571428571, 2.34642857142857, 2.01746031746032]^T$. The correct answer is $x = [1, 2, 3, 4, 5]^T$.

6. Write up your own *LU* factorization code, based on the material in this chapter, and test it on the following examples. In each case have your code multiply out the L and U factors to check that the routine is working.

   (a) $K_5 x = b$, $b = [-1, 0, 0, 0, 5]^T$; the solution is $x = [0, 1, 2, 3, 4]^T$;
   (b) $A_5 x = b$, $b = [-4, -7, -6, -5, 16]^T$; the solution is $x = [0, 1, 2, 3, 4]^T$.

Solution: For (a) you get

$$
L = \begin{bmatrix}
1.0000 & 0 & 0 & 0 & 0 \\
-0.5000 & 1.0000 & 0 & 0 & 0 \\
0 & -0.6667 & 1.0000 & 0 & 0 \\
0 & 0 & -0.7500 & 1.0000 & 0 \\
0 & 0 & 0 & -0.8000 & 1.0000
\end{bmatrix}
$$

and

$$
U = \begin{bmatrix}
2.0000 & -1.0000 & 0 & 0 & 0 \\
0 & 1.5000 & -1.0000 & 0 & 0 \\
0 & 0 & 1.3333 & -1.0000 & 0 \\
0 & 0 & 0 & 1.2500 & -1.0000 \\
0 & 0 & 0 & 0 & 1.2000
\end{bmatrix}.
$$

7. Determine the operation count for computing the inverse of a matrix, as outlined in this section.

Solution: If C_{GE} is the cost of naive Gaussian elimination, and C_s is the cost of the backsolve steps, then the obvious cost of computing the inverse is

$$
C_I = C_{GE} + nC_s = \frac{1}{3}n^3 + \mathcal{O}(n^2) + \frac{1}{2}n^3 + \mathcal{O}(n^2) = \frac{5}{6}n^3 + \mathcal{O}(n^2).
$$

But there are extra costs associated with the elimination process (because the right side now has n vectors instead of only 1) and this is where the figure of $(4/3)n^3 + \mathcal{O}(n^2)$ comes from. Some costs can be saved (roughly $(1/3)n^3$ operations) by carefully taking advantage of the fact that the right side vectors are all initially ones or zeroes.

8. Show that

(a)

$$
E_k^{-1} = I + R_k \tag{7.1}
$$

for all k;

Solution:

$$
(I + R_k)(I - R_k) = I + R_k - R_k - R_k^2 = I - R_k^2.
$$

But $R_k^2 = 0$, because of the special position of the non-zero elements, thus $E_k^{-1} = I + R_k$.

(b)

$$
E_k^{-1} E_{k-1}^{-1} = I + R_k + R_{k-1} \tag{7.2}
$$

for all k;

Solution:

$$
(I + R_k)(I + R_{k-1}) = I + R_k + R_{k-1} + R_k R_{k-1}.
$$

Again, the cross product term is zero because of the placement of the non-zero terms.

(c) $L = I + R_1 + R_2 + \cdots + R_{n-1}$.

Solution: This follows from the two previous parts.

9. Modify the tridiagonal solution algorithm from Chapter 2 to produce an *LU* decomposition. Be sure to maintain the simple storage of the matrix that was used in Chapter 2, and assume that no pivoting is required.

   Solution: A MATLAB script for doing this follows.

```
function [l1, d1, u1] = trifact(l,d,u)
    n = length(d);
    u1 = u;
    d1 = d;
    l1 = u1;
    l1(1) = 0;
    for k=2:n
        l1(k) = l(k)/d1(k-1);
        d1(k) = d(k) - u(k-1)*l1(k)/d1(k-1);
    end
```

10. Write an *LU* factorization code with partial pivoting, and apply it to the system $A_5 x = b$, where $b = [-4, -7, -6, -5, 16]^T$ and the solution is $x = [0, 1, 2, 3, 4]^T$.

    Solution: We get the factors

    $$L = \begin{bmatrix} 1 & 0 & 0 & 0 & 0 \\ 0.25 & 1 & 0 & 0 & 0 \\ 0 & 0 & 1 & 0 & 0 \\ 0 & -0.9412 & 0.4853 & 1 & 0 \\ 0 & 0 & 0 & -0.8918 & 1 \end{bmatrix}$$

    and

    $$U = \begin{bmatrix} 4 & 1 & -4 & 0 & 0 \\ 0 & -4.25 & 1 & 0 & 0 \\ 0 & 0 & 4 & 1 & -4 \\ 0 & 0 & 0 & -4.4853 & 1.9412 \\ 0 & 0 & 0 & 0 & 2.7311 \end{bmatrix}$$

    so the product is

    $$A' = LU = \begin{bmatrix} 4 & 1 & -4 & 0 & 0 \\ 1 & -4 & 0 & 0 & 0 \\ 0 & 0 & 4 & 1 & -4 \\ 0 & 4 & 1 & -4 & 0 \\ 0 & 0 & 0 & 4 & 1 \end{bmatrix}.$$

11. Write an *LU* factorization code that does partial pivoting and use it to solve the system of equations $Ax = b$ where

    $$A = \begin{bmatrix} 9 & 3 & 2 & 0 & 7 \\ 7 & 6 & 9 & 6 & 4 \\ 2 & 7 & 7 & 8 & 2 \\ 0 & 9 & 7 & 2 & 2 \\ 7 & 3 & 6 & 4 & 3 \end{bmatrix}$$

    and $b = [35, 58, 53, 37, 39]^T$. The correct answer is $x = [0, 1, 2, 3, 4]^T$.

Solution: This is a simple implementation of the pseudocode in the text. We get the factors

$$L = \begin{bmatrix} 1 & 0 & 0 & 0 & 0 \\ 0 & 1 & 0 & 0 & 0 \\ 0.7778 & 0.4074 & 1 & 0 & 0 \\ 0.2222 & 0.7037 & 0.3548 & 1 & 0 \\ 0.7778 & 0.0741 & 0.8548 & -0.1222 & 1 \end{bmatrix}$$

and

$$U = \begin{bmatrix} 9 & 3 & 2 & 0 & 7 \\ 0 & 9 & 7 & 2 & 2 \\ 0 & 0 & 4.5926 & 5.1852 & -2.2593 \\ 0 & 0 & 0 & 4.7527 & -0.1613 \\ 0 & 0 & 0 & 0 & -0.6810 \end{bmatrix}$$

so the product is

$$A' = LU = \begin{bmatrix} 9 & 3 & 2 & 0 & 7 \\ 0 & 9 & 7 & 2 & 2 \\ 7 & 6 & 9 & 6 & 4 \\ 2 & 7 & 7 & 8 & 2 \\ 7 & 3 & 6 & 4 & 3 \end{bmatrix},$$

which, as expected, is the same as the original matrix with the rows re-ordered.

12. Compare your LU factorization-and-solution code to MATLAB'S linsolve command by creating a random 10×10 system and solving with both routines. (They should produce *exactly* the same solution.)

13. Again, generate a random 10×10 linear system (matrix and right-hand side vector). Then solve this system four ways:

    (a) Using your LU factorization-and-solution code;

    (b) Using MATLAB's linsolve command;

    (c) Using the MATLAB backslash operation;

    (d) Using MATLAB's inv command to compute the inverse of the matrix, and then multiply this by the right-hand side vector.

    Use timing routines to estimate the cost of each solution technique, and rank the methods for their efficiency in this regard.

14. Repeat Problem 13 for the matrix K_{20}, defined at the end of §7.2, using a random right-hand side vector. Then apply the tridiagonal solver from Chapter 2 to this problem, and again get the timing estimates. Comment on your results.

15. Let's return to Algorithm 2.6; rewrite this to take a symmetric tridiagonal matrix, T, defined only by the two diagonal vectors da and la, and return the factors L and D, such that $LDL^T = T$, where D is diagonal and L is unit lower diagonal.

◁ • • ▷

7.5 PERTURBATION, CONDITIONING AND STABILITY

Exercises:

1. Let

$$A = \begin{bmatrix} 1 & 2 & -7 \\ 4 & -8 & 0 \\ 2 & 1 & 0 \end{bmatrix}.$$

Compute $\|A\|_\infty$.

Solution: $\|A\|_\infty = \max\{10, 12, 3\} = 12.$

2. Let

$$A = \begin{bmatrix} 5 & 6 & -9 \\ 1 & 2 & 3 \\ 0 & 7 & 2 \end{bmatrix}.$$

Compute $\|A\|_\infty$.

Solution: $\|A\|_\infty = \max\{20, 6, 9\} = 20.$

3. Let

$$A = \begin{bmatrix} -8 & 0 & -1 \\ 3 & 12 & 0 \\ 1 & 2 & 3 \end{bmatrix}.$$

Compute $\|A\|_\infty$.

Solution: $\|A\|_\infty = \max\{9, 15, 6\} = 15.$

4. Show that

$$\|A\|_\star = \max_{i,j} |a_{ij}|$$

does not define a matrix norm, according to our defintion. Hint: show that one of the conditions fails to hold by finding a specific case where it fails.

Solution: Let

$$A = B = \begin{bmatrix} 1 & 1 \\ 1 & 1 \end{bmatrix}.$$

Then

$$AB = \begin{bmatrix} 2 & 2 \\ 2 & 2 \end{bmatrix},$$

so $\|AB\|_\star = 2$, but $\|A\|_\star = \|B\|_\star = 1$ so the inequality $\|AB\|_\star \leq \|A\|_\star \|B\|_\star$ doesn't hold.

5. Let

$$A = \begin{bmatrix} 1 & \frac{1}{2} & 0 \\ \frac{1}{2} & \frac{1}{3} & \frac{1}{4} \\ 0 & \frac{1}{4} & \frac{1}{5} \end{bmatrix}.$$

Compute, directly from the definition, $\kappa_\infty(A)$. You should get $\kappa_\infty(A) = 18$.

Solution: We get

$$A^{-1} = \frac{1}{11} \begin{bmatrix} -1 & 24 & -30 \\ 24 & -48 & 60 \\ -30 & 60 & -20 \end{bmatrix},$$

so $\|A^{-1}\|_\infty = 132/11 = 12$. Since $\|A\|_\infty = 3/2$, we get the expected result.

6. Repeat the above for

$$A = \begin{bmatrix} 4 & 1 & 0 \\ 1 & 4 & 1 \\ 0 & 1 & 4 \end{bmatrix},$$

for which $\kappa_\infty(A) = 2.5714$.

Solution: We get

$$A^{-1} = \frac{1}{56} \begin{bmatrix} 15 & -4 & 1 \\ -4 & 16 & -4 \\ 1 & -4 & 15 \end{bmatrix},$$

so $\|A^{-1}\|_\infty = 3/7$. Since $\|A\|_\infty = 6$, we get the expected result.

7. Consider the linear system

$$\begin{bmatrix} 1.002 & 1 \\ 1 & 0.998 \end{bmatrix} \begin{bmatrix} x_1 \\ x_2 \end{bmatrix} = \begin{bmatrix} 0.002 \\ 0.002 \end{bmatrix},$$

which has the exact solution $x = (1, -1)^T$ (verify this). What is the residual $b - Ax_c$ for the "approximate" solution $x_c = (0.29360067817338, -0.29218646673249)^T$. Explain.

Solution: We have $r = 10^{-5} \times (-0.1413, 0.1416)^T$. Note that an "approximate solution" which is not very close to the actual solution produced a very small residual, thus demonstrating that a small residual does not always mean that the computed solution is close to accurate.

8. Consider the linear system problem

$$Ax = b,$$

where

$$A = \begin{bmatrix} 4 & 2 & 0 \\ 1 & 4 & 1 \\ 0 & 2 & 4 \end{bmatrix}, \quad b = \begin{bmatrix} 8 \\ 12 \\ 16 \end{bmatrix},$$

for which the exact solution is $x = (1, 2, 3)^T$. (Check this.) Let $x_c = x + (0.002, 0.01, 0.001)^T$ be a computed (i.e., approximate) solution to this system. Use Theorem 7.8 to find the perturbation matrix E such that x_c is the exact solution to $Ax_c = b$.

Solution: We get

$$E = \begin{bmatrix} -0.0020 & -0.0040 & -0.0060 \\ -0.0031 & -0.0062 & -0.0092 \\ -0.0017 & -0.0034 & -0.0051 \end{bmatrix}.$$

9. Compute the growth factor for Gaussian elimination for the matrix in the previous problem.

   Solution: We get

   $$A^{(1)} = \begin{bmatrix} 4 & 2 & 0 \\ 0 & 3.5 & 1 \\ 0 & 2 & 4 \end{bmatrix},$$

   $$A^{(2)} = \begin{bmatrix} 4 & 2 & 0 \\ 0 & 3.5 & 1 \\ 0 & 0 & 3.4286 \end{bmatrix},$$

   so $\rho = 2/3$.

10. Let

    $$A = \begin{bmatrix} 2 & 1 & 0 \\ 1 & 2 & 1 \\ 0 & 1 & 2 \end{bmatrix}.$$

    This has exact condition number $\kappa_\infty(A) = 8$. Use the condition number estimator in this section to approximate the condition number.

    Solution: $\kappa^* = 7.1178$. This value will of course be affected by different random initial guesses for the iteration.

11. Repeat the above for

    $$A = \begin{bmatrix} 1 & \frac{1}{2} & 0 \\ \frac{1}{2} & \frac{1}{3} & \frac{1}{4} \\ 0 & \frac{1}{4} & \frac{1}{5} \end{bmatrix}$$

    for which $\kappa_\infty(A) = 18$.

    Solution: $\kappa^* = 13.3186$.

12. Use the condition number estimator to produce a plot of κ^* versus n for each of the following matrix families:

    (a) $T_n, 4 \leq n \leq 20$;

    (b) $K_n, 4 \leq n \leq 20$;

    (c) $H_n, 4 \leq n \leq 20$.

    (d) $A_n, 4 \leq n \leq 20$.

    Compare your estimates with the exact values from cond and the estimates from rcond.

    Solution: Figure 7.1 shows the plots. Note that the scales are very different, and that the H_n matrices have enormous condition numbers for n as small as 10.

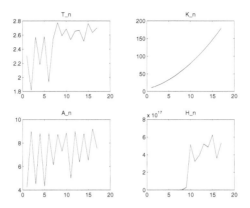

Figure 7.1   Solution plot for Exercise 7.5.12.

13. Produce a plot of the growth factor for Gaussian elimination for each matrix family in the above problem, as a function of n.

Solution: Figure 7.2 shows the plots.

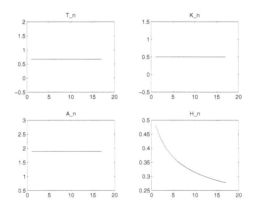

Figure 7.2   Solution plot for Exercise 7.5.13.

14. Given a matrix $A \in \mathbb{R}^{n \times n}$, show that

$$\mu(A) = \frac{\max_{x \neq 0} \frac{\|Ax\|}{\|x\|}}{\min_{x \neq 0} \frac{\|Ax\|}{\|x\|}}$$

is equivalent to the condition number as defined in (7.10).

Solution: We have

$$\mu(A) = \frac{\max_{x \neq 0} \frac{\|Ax\|}{\|x\|}}{\min_{x \neq 0} \frac{\|Ax\|}{\|x\|}} = \frac{\|A\|}{\min_{x \neq 0} \frac{\|Ax\|}{\|x\|}}$$

so it remains only to show that the denominator is equivalent to $1/\|A^{-1}\|$. Since

$$\frac{1}{\min X} = \max \frac{1}{X},$$

we have

$$\frac{1}{\min_{x \neq 0} \frac{\|Ax\|}{\|x\|}} = \max_{x \neq 0} \frac{\|x\|}{\|Ax\|} = \max_{y \neq 0} \frac{\|A^{-1}y\|}{\|y\|} = \|A^{-1}\|$$

and we are done.

15. Prove that (7.8) and (7.9) follow from the definition of matrix norm.

    Solution: We have that

    $$\|A\| = \max x \neq 0 \frac{\|Ax\|}{\|x\|}$$

    so that, for any particular choice of x,

    $$\|A\| \geq \frac{\|Ax\|}{\|x\|},$$

    which proves (7.9). To get (7.8), we simply apply (7.9):

    $$\|AB\| = \max x \neq 0 \frac{\|ABx\|}{\|x\|} \leq \max x \neq 0 \frac{\|A\|\|Bx\|}{\|x\|} = \|A\|\|B\|.$$

16. Prove Theorem 7.10.

    Solution: Theorem 7.8 says that there exists a perturbation matrix E such that x_c is the exact solution to $(A + E)x_c = b$. Theorem 7.6 says that if $\kappa(A)\|E\| \leq \|A\|$, then the relative error is bounded above according to

    $$\frac{\|x - x_x\|}{\|x\|} \leq \frac{\theta}{1 - \theta},$$

    where

    $$\theta = \kappa \frac{\|E\|}{\|A\|}.$$

    The function $f(\theta) = \theta/(1 - \theta)$ is an increasing function, so replacing $\kappa \frac{\|E\|}{\|A\|}$ with something larger preserves the inequality. Returning to Theorem 7.8, we have that

    $$\frac{\|E\|_2}{\|A\|_2} \leq \frac{\|r\|_2}{\|A\|_2\|x_c\|_2} \leq \frac{\|r\|_2}{\|Ax_c\|_2}.$$

    But

    $$b - r = b - (b - Ax_c) = Ax_c,$$

    so we are done.

17. Give an argument to support the validity of the "rule of thumb" following Theorem 7.10.

    Solution: Follows by applying the given assumptions to Theorem 7.6. We have that

    $$\frac{\|x - x_x\|}{\|x\|} \leq \frac{\theta}{1 - \theta},$$

where

$$\theta = \kappa \frac{\|E\|}{\|A\|},$$

but

$$\theta \leq (C_4 \times 10^t)(C_1 C_2 \times 10^{-s}) = C \times 10^{t-s}.$$

18. Prove Theorem 7.5.

Solution: We have $r = b - Ax_c = Ax - Ax_c$ so $x - x_c = A^{-1}r$ and the result follows along the same lines as Theorem 7.4.

19. Consider the linear system problem $Ax = b$ for

$$A = \begin{bmatrix} 1 & \frac{1}{2} \\ \frac{1}{2} & \frac{1}{3} \end{bmatrix}, \quad b = \begin{bmatrix} 2 \\ 7/6 \end{bmatrix}.$$

Note that the exact solution is $x = (1, 2)^T$.

(a) Represent the exact factorization of A using only four decimal digits, and solve the system.

(b) Do two steps of iterative refinement to improve your solution.

Solution: The factorization is

$$L = \begin{bmatrix} 1.0000 & 0 \\ 0.5000 & 1.0000 \end{bmatrix},$$

$$U = \begin{bmatrix} 1.0000 & 0.5000 \\ 0 & 0.0833 \end{bmatrix},$$

which multiplies out to

$$LU = \begin{bmatrix} 1.0000 & 0.5000 \\ 0.5000 & 0.3333 \end{bmatrix}.$$

The uncorrected solution is

$$x_c = \begin{pmatrix} 0.9996 \\ 2.0008 \end{pmatrix}$$

and the refined solution (after two iterations) is

$$x = \begin{pmatrix} 0.99999999993592 \\ 2.00000000012816 \end{pmatrix}.$$

20. Do two more steps of refinement for the problem in Example 7.9.

Solution: $x^{(3)} = (0.99999999999999, 2.00000000000004, 2.99999999999996)^T.$

◁ • • ▷

7.6 SPD MATRICES AND THE CHOLESKY DECOMPOSITION

Exercises:

1. Show that a matrix with one or more non-positive diagonal elements cannot be SPD. Hint: Try to find a vector x such that $x^T A x = a_{ii}$.

   Solution: If a_{ii} is the non-positive diagonal element, take $x = e_i$. Then $e_i^T A e_i = a_{ii}$.

2. For a 2×2 SPD matrix, show that $a_{22} - a_{21}^2/a_{11} > 0$. Hint: Consider the positive definite condition with $x = (a_{21}, -a_{11})^T$.

   Solution: For x as suggested, we have

   $$x^T A x = a_{11}^2 a_{22} - a_{11} a_{21}^2 = a_{11}\left(a_{22} - a_{21}^2/a_{11}\right).$$

3. Let $A \in \mathbb{R}^{n \times n}$ be partitioned as

   $$A = \begin{bmatrix} A_{11} & A_{21} \\ A_{21} & A_{22} \end{bmatrix},$$

   where A_{11} and A_{22} are square matrices. Show that both A_{11} and A_{22} must be SPD if $A \in SPD$.

   Solution: Suppose A_{11} is not SPD. Then take $x = (y, 0)^T$, partitioned conformably with A, and with y such that $y^T A_{11} y \leq 0$. Then, since

   $$x^T A x = y^T A_{11} y,$$

   A cannot be SPD.

4. In the proof of the Cholesky theorem, show that $a_{kk} - g^T g > 0$ must hold, or else there is a vector x such that $x^T A x \leq 0$. Hint: Use Problem 2 as a starting point.

   Solution: Follows from the hint and Exercise 1.

5. Derive the operation count given in the text for the Cholesky decomposition.

   Solution: Follows from the formulas in the text and a direct computation.

6. Prove the following: If $A \in SPD$, then there exists a unit lower triangular matrix L, and a diagonal matrix D with positive elements, such that $A = LDL^T$.

   Solution: Let $D_1 = \text{diag}(G)$, where G is the usual Cholesky factor. Define $L = GD_1^{-1}$ and $D = D_1^2$. Then

   $$LDL^T = (GD_1^{-1})D_1^2(D_1^{-1}G^T) = GG^T = A.$$

7. Derive an algorithm for the LDL^T factorization that does not require any square roots. Hint: Look at the 3×3 case, explicitly. Multiply out LDL^T and set the result equal to A, and from this deduce the relationships that are necessary. In other words, if

   $$\begin{bmatrix} 1 & 0 & 0 \\ l_{21} & 1 & 0 \\ l_{31} & l_{32} & 1 \end{bmatrix} \begin{bmatrix} d_{11} & 0 & 0 \\ 0 & d_{22} & 0 \\ 0 & 0 & d_{33} \end{bmatrix} \begin{bmatrix} 1 & l_{21} & l_{31} \\ 0 & 1 & l_{32} \\ 0 & 0 & 1 \end{bmatrix} = \begin{bmatrix} a_{11} & a_{12} & a_{13} \\ a_{21} & a_{22} & a_{23} \\ a_{31} & a_{32} & a_{33} \end{bmatrix}$$

what is the relationship between the components of L, D, and A? Then generalize to a problem of arbitrary size. Don't forget that A is symmetric!

8. The T_n and K_n families of matrices are both positive definite. Write up a code to do the Cholesky factorization and test it on these matrices over the range $3 \leq n \leq 20$. Use a right-side vector that is all ones, and confirm that the code is working by computing $\|r\|_\infty$, where r is the residual vector.

Solution: For both families, the norm of the residual is on the order of 10^{-12} or smaller, for the entire range of problem sizes given here.

9. If A is tridiagonal *and* SPD, then the Cholesky factorization can be modified to work only with the *two* distinct non-zero diagonals in the matrix. Construct this version of the algorithm and test it on the T_n and K_n families, as above.

Solution: A MATLAB script that does this is given below.

```
function [l,d] = chol3(l,d)
    n = length(d);
    d(1) = sqrt(d(1));
    for k=2:n
        l(k) = l(k)/d(k-1);
        d(k) = sqrt(d(k) - l(k)^2);
    end
```

10. The H_n family of matrices is also SPD, but ill-conditioned. Try to apply chol, over the range $4 \leq n \leq 20$. What happens? Can you explain this?

Solution: For $n \leq 13$ everything is fine, but for $n = 14$ the routine reports that the matrix is no longer SPD. In terms of the computation, the argument to the square root in the computation of g_{ii} becomes negative.

◁ • • • ▷

7.7   APPLICATION: NUMERICAL SOLUTION OF LINEAR LEAST SQUARES PROBLEMS

Exercises:

1. Use the normal equations to solve (in the least squares sense) each of the following linear systems. Compute the residual in each case.

(a) $\begin{bmatrix} 1 & 2 \\ 3 & 4 \\ 5 & 6 \end{bmatrix} \begin{bmatrix} x_1 \\ x_2 \end{bmatrix} = \begin{bmatrix} 1 \\ 2 \\ 3 \end{bmatrix}$;

(b) $\begin{bmatrix} 4 & 1 \\ 1 & 4 \\ 0 & 1 \end{bmatrix} \begin{bmatrix} x_1 \\ x_2 \end{bmatrix} = \begin{bmatrix} 1 \\ 2 \\ 3 \end{bmatrix}$;

(c) $\begin{bmatrix} 2 & -1 \\ -1 & 2 \\ 0 & -1 \end{bmatrix} \begin{bmatrix} x_1 \\ x_2 \end{bmatrix} = \begin{bmatrix} 1 \\ 4 \\ 1 \end{bmatrix}$;

(d) $\begin{bmatrix} 4 & 1 & 0 \\ 1 & 4 & 1 \\ 0 & 1 & 4 \\ 1 & 1 & 1 \end{bmatrix} \begin{bmatrix} x_1 \\ x_2 \\ x_3 \end{bmatrix} = \begin{bmatrix} 1 \\ 2 \\ 2 \\ 1 \end{bmatrix}.$

Solution: (a) We have

$$B = A^T A = \begin{bmatrix} 35 & 44 \\ 44 & 56 \end{bmatrix}$$

and

$$f = A^T b = \begin{bmatrix} 22 \\ 28 \end{bmatrix},$$

so the normal equations solution is

$$x = \begin{bmatrix} 0 \\ 0.5 \end{bmatrix},$$

which is exact, i.e., the residual $r = b - Ax$ is zero. This means that $b = \begin{bmatrix} 1 \\ 2 \\ 3 \end{bmatrix}$ is in

the *column space* of A.

Solution: (d) We now have

$$B = A^T A = \begin{bmatrix} 18 & 9 & 2 \\ 9 & 19 & 9 \\ 2 & 9 & 18 \end{bmatrix}$$

and

$$f = A^T b = \begin{bmatrix} 7 \\ 12 \\ 11 \end{bmatrix},$$

so the normal equations solution is

$$x = \begin{bmatrix} 0.1640 \\ 0.3578 \\ 0.4140 \end{bmatrix}$$

and the residual is now

$$r = b - Ax = \begin{bmatrix} -0.0138 \\ -0.0092 \\ -0.0138 \\ 0.0642 \end{bmatrix},$$

with $\|x\|_2 = 0.0677$.

2. Use a QR factorization to solve (in the least squares sense) each of the following linear systems. Use the native qr routine in whatever computing environment you wish. Compute the residual in each case.

(a)

$$
\begin{bmatrix}
-1 & 1 \\
-0.8 & 1 \\
-0.6 & 1 \\
-0.4 & 1 \\
-0.2 & 1 \\
0 & 1 \\
0.2 & 1 \\
0.4 & 1 \\
0.6 & 1 \\
0.8 & 1 \\
1 & 1
\end{bmatrix}
\begin{bmatrix} m \\ b \end{bmatrix}
=
\begin{bmatrix}
-2.97287 \\
-2.57903 \\
-2.19701 \\
-1.79938 \\
-1.37186 \\
-0.95499 \\
-0.57362 \\
-0.18853 \\
0.21269 \\
0.64054 \\
1.02317
\end{bmatrix}
;
$$

(b)

$$
\begin{bmatrix}
1 & \frac{1}{2} & \frac{1}{3} \\
0.25 & 0.2 & \frac{1}{6} \\
\frac{1}{7} & 0.125 & \frac{1}{9} \\
0.1 & \frac{1}{11} & \frac{1}{12} \\
\frac{1}{13} & \frac{1}{14} & \frac{1}{15}
\end{bmatrix}
\begin{bmatrix} x_1 \\ x_2 \\ x_3 \end{bmatrix}
=
\begin{bmatrix} 1 \\ 2 \\ 3 \\ 4 \end{bmatrix}
;
$$

(c)

$$
\begin{bmatrix}
0.25 & -0.5 & 1 \\
0.16 & -0.4 & 1 \\
0.09 & -0.3 & 1 \\
0.04 & -0.2 & 1 \\
0.01 & -0.1 & 1 \\
0 & 0 & 1 \\
0.01 & 0 & 1 \\
0.04 & 0.2 & 1 \\
0.09 & 0.3 & 1 \\
0.16 & 0.4 & 1 \\
0.25 & 0.5 & 1
\end{bmatrix}
\begin{bmatrix} x_1 \\ x_2 \\ x_3 \end{bmatrix}
=
\begin{bmatrix}
0.3612 \\
1.1403 \\
0.8797 \\
0.8817 \\
1.2139 \\
1.0965 \\
1.3420 \\
2.3821 \\
2.6461 \\
2.5352 \\
2.3098
\end{bmatrix}
;
$$

(d)

$$
\begin{bmatrix}
1 & 1 \\
0.9 & 1.1 \\
0.99 & 1.01 \\
0.999 & 1.001
\end{bmatrix}
\begin{bmatrix} x_1 \\ x_2 \end{bmatrix}
=
\begin{bmatrix} 1 \\ 2 \\ 2 \\ 1 \end{bmatrix}.
$$

Solution: (d) For $A = \begin{bmatrix} 1 & 1 \\ 0.9 & 1.1 \\ 0.99 & 1.01 \\ 0.999 & 1.001 \end{bmatrix}$, MATLAB's qr command returns

$$
Q =
\begin{bmatrix}
-0.5138 & 0.3094 & -0.5423 & -0.5884 \\
-0.4624 & -0.8830 & -0.0795 & 0.0128 \\
-0.5087 & 0.1901 & 0.8140 & -0.2061 \\
-0.5133 & 0.2974 & -0.1923 & 0.7817
\end{bmatrix}
$$

and

$$R' = \begin{bmatrix} -1.9463 & -2.0500 \\ 0 & -0.1722 \\ 0 & 0 \\ 0 & 0 \end{bmatrix}$$

so that

$$R = \begin{bmatrix} -1.9463 & -2.0500 \\ 0 & -0.1722 \end{bmatrix}$$

and

$$f = Q^T b = \begin{bmatrix} -2.9692 \\ -0.7790 \\ 0.7345 \\ -0.1933 \end{bmatrix}.$$

We then solve the system $Rx = \begin{bmatrix} -2.9692 \\ -0.7790 \end{bmatrix}$ to get $x = \begin{bmatrix} -3.2391 \\ 4.5236 \end{bmatrix}$. The residual is

$$r = b - Ax = \begin{bmatrix} -0.2846 \\ -0.0609 \\ 0.6378 \\ -0.2923 \end{bmatrix},$$

and $\|r\|_2 = 0.7596$.

3. Pick one example from each of the two previous exercises (your choice). Use the rand function to generate random vectors of the same length as the least squares solution, and confirm that the least squares residual is, in fact, smaller than any of these other residuals.

Solution: If we pick Problem 2(d) (the immediately previous exercise) and write a script to chose, say, 5 random vectors in $\mathbb{R}^2$, we get something like this:

Table 7.1   Random vector tests (Problem 3), using Problem 2(d).

k	$x^{(k)}$	$r^{(k)} = b - Ax^{(k)}$	$\|r^{(k)}\|_2$
1	$(0.1576, 0.9706)^T$	$(-0.1282, 0.7905, 0.8637, -0.1290)^T$	1.1849
2	$(0.9572, 0.4854)^T$	$(-0.4425, 0.6046, 0.5622, -0.4421)^T$	1.0358
3	$(0.8003, 0.1419)^T$	$(0.0578, 1.1237, 1.0644, 0.0585)^T$	1.5500
4	$(0.4218, 0.9157)^T$	$(-0.3375, 0.6131, 0.6576, -0.3380)^T$	1.0181
5	$(0.7922, 0.9595)^T$	$(-0.7517, 0.2316, 0.2466, -0.7519)^T$	1.1157

Note that all of the residuals are indeed larger than $\|r\|_2 = 0.7596$.

4. Show that the formulas given for the slope m and y-intercept b for a least squares data fit to a straight line (Eq'ns (4.56) and (4.57) in §4.10.1) are precisely the solutions to a normal equations problem.

Solution: If we take the standard equation for a straight line, $y = mx + b$, and try to compute the pair (m, b) by fitting n data points $(x_k, y_k)_{k=1}^{k=n}$ to this equation, we get something like this:

$$
\begin{bmatrix} x_1 & 1 \\ x_2 & 1 \\ \vdots & \\ x_n & 1 \end{bmatrix} \begin{bmatrix} m \\ b \end{bmatrix} = \begin{bmatrix} y_1 \\ y_2 \\ \vdots \\ y_n \end{bmatrix}. \tag{7.3}
$$

(If we have a lot of data points, this is a classic case of a "skinny" least squares problem, in that the matrix is much "taller" than it is "wide.") We can re-write this as a least squares problem,

$$
A z_{m,b} = b_y,
$$

where A is the matrix in (7.3), and b_y is the right-hand side. Using the normal equations approach means we must solve

$$
A^T A z_{m,b} = A^T b_y,
$$

and a little algebraic manipulation shows that this is equivalent to (4.56) and (4.57) in §4.10.1.

5. Let $B = A^T A \in \text{SPD}$ for a full-rank matrix $A \in \mathbb{R}^{n \times n}$, and let R be the upper triangular Cholesky factor of B. Define $Q = (RA^{-1})^T$, and prove that $Q^T Q = I$. (How do we know that A^{-1} exists?)

6. Let $Q \in \mathbb{R}^{n \times n}$ be an orthogonal matrix; show that $\|Qx\|_2 = \|x\|_2$ for any $x \in \mathbb{R}^n$.

7. The data in Table 7.2 is supposed to represent a relation of the form $E = CN^{-r}$, for some r. By taking logarithms, set this up as a linear least-squares problem and find an estimate for r. Plot the data and resulting curve.

Table 7.2   Data for Problem 7.

N	8	12	16	24	32	48
E	5.24×10^{-5}	1.37×10^{-5}	4.73×10^{-6}	1.84×10^{-6}	9.21×10^{-7}	3.83×10^{-7}

◁ • • ▷

7.8   SPARSE AND STRUCTURED MATRICES

(No exercises in this section.)

◁ • • ▷

7.9 ITERATIVE METHODS FOR LINEAR SYSTEMS – A BRIEF SURVEY

Exercises:

1. Let

$$A = \begin{bmatrix} 4 & -1 & 0 & 0 \\ -1 & 4 & -1 & 0 \\ 0 & -1 & 4 & -1 \\ -1 & 0 & -1 & 4 \end{bmatrix}$$

and $b = (-4, 2, 4, 10)^T$.

(a) Verify that the solution to $Ax = b$ is $x = (0, 1, 2, 3)^T$.

(b) Do three iterations (by hand) of the Jacobi iteration for this matrix, using $x^{(0)} = (0, 0, 0, 0)^T$.
   Solution: $x^{(3)} = (-0.0781, 0.9062, 1.8438, 2.9062)^T$.

(c) Do three iterations (by hand) of the Gauss-Seidel iteration for this problem, using the same initial guess.
   Solution: $x^{(3)} = (-0.0645, 0.9500, 1.9702, 2.9764)^T$.

2. Do three iterations of SOR for the previous example, using $\omega = 1.05$.
   Solution: $x^{(3)} = (-0.0959, 0.9430, 1.6624, 2.9908)^T$

3. Solve the same system using SOR for a wide set of values of $\omega \in (0, 2)$. For each ω, compute $r(\omega) = \|b - Ax^{(k)}\|_\infty$. Graph $r(\omega)$ for $k = 1, 3, 5, 10$.

   Solution: For $\omega = 0.25, 0.50, 0.75, 1.00, 1.25, 1.50$, and 1.75 we get the plots in Figure 7.3. Note that the residual gets very large away from the critical value of ω, which appears to be slightly larger than 1.

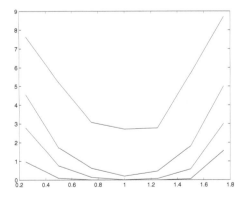

Figure 7.3   Solution plots for Exercise 7.7.3.

4. Write a computer code that does Jacobi for Problem 1, for a specified number of iterations. How many iterations does it take to get convergence, in the sense that the consecutive iterates differ by less than 10^{-6}?

Solution: Convergence occurs in 18 iterations.

5. Repeat the above for Gauss-Seidel.

Solution: Convergence occurs in 12 iterations.

6. Repeat the above for SOR. Make sure your code can accept different values of ω as an input parameter.

Solution: For $\omega = 1.25$, convergence occurs in 16 iterations; note that this is better than Jacobi, but worse than Gauss-Seidel. However, for $\omega = 1.05$, we get convergence in 11 iterations, which is better than Gauss-Seidel.

7. Let A be the 16×16 matrix given at the beginning of this section. Take

$$b = (5, 11, 18, 21, 29, 40, 48, 48, 57, 72, 80, 76, 69, 87, 94, 85)^T.$$

Write a computer code to do Jacobi, Gauss-Seidel, and SOR on this system of equations. Write the code to only store the non-zero diagonals of A, and make the code as efficient as possible.

Solution: Jacobi converges in about 80 iterations, Gauss-Seidel in about 43, and SOR (with $\omega = 1.25$) in about 20.

8. Prove that the spectral radius of A, $\rho(A)$, is bounded above by $\|A\|$ for any norm such that $\|Ax\| \leq \|A\|\|x\|$.

Solution: Let λ be the eigenvalue which generates the spectral radius. Then $\lambda x = Ax$ and so

$$\|\lambda x\| = \|Ax\| \Rightarrow \rho = \frac{\|Ax\|}{\|x\|} \leq \|A\|.$$

◁ • • • ▷

7.10 NONLINEAR SYSTEMS: NEWTON'S METHOD AND RELATED IDEAS

Exercises:

1. Consider the nonlinear system

$$2x_1 - x_2 + \frac{1}{9}e^{-x_1} = -1, \tag{7.4}$$

$$-x_1 + 2x_2 + \frac{1}{9}e^{-x_2} = 1. \tag{7.5}$$

Take $x^{(0)} = (1, 1)^T$ and do two iterations of Newton's method; you should get

$$x^{(2)} = (-0.48309783661427, 0.21361449746996)^T.$$

Solution: This is fairly straight-forward. As an intermediate check, you should get

$$x^{(1)} = (-0.42317280134882, 0.25270278130481)^T.$$

2. Write a computer code to solve the system in the preceding problem, using

    (a) Newton's method;

    (b) The chord method;

    (c) The chord method, updating every 3 iterations.

Solution: Below is a (very inefficient) MATLAB script for the Newton iteration.

```
function z = newt2(x1,x2,n)
    for k=1:n
        y1 = 2*x1 - x2 + (1/9)*exp(-x1) + 1;
        y2 = -x1 + 2*x2 + (1/9)*exp(-x2) - 1;
        df = [2 - exp(-x1)/9 -1; -1 2 - exp(-x2)/9];
        dfi = inv(df);
        z = [x1 x2]' - dfi*[y1 y2]'
        x1 = z(1);
        x2 = z(2);
    end
```

3. Re-write the system in Problem 1 as

$$Kx + \phi(x) = b,$$

where K is the 2×2 matrix

$$K = \begin{bmatrix} 2 & -1 \\ -1 & 2 \end{bmatrix},$$

$\phi(x)$ is defined by

$$\phi(x) = \begin{pmatrix} \frac{1}{9}e^{-x_1} \\ \frac{1}{9}e^{-x_2} \end{pmatrix},$$

and $b = (-1, 1)^T$.

    (a) Do two iterations (by hand) of the fixed-point iteration

$$x^{(k+1)} = \frac{1}{2}(b - \phi(x^{(k)}) - Kx^{(k)} + 2x^{(k)})$$

    for this system, using $x^{(0)} = (1, 1)^T$.
    Solution:

$$x^{(1)} = (-0.02043774673175, 0.97956225326825)^T,$$

$$x^{(2)} = (-0.06692154156532, 0.46892138074399)^T.$$

    (b) Do two iterations (by hand) of the fixed-point iteration

$$x^{(k+1)} = K^{-1}(b - \phi(x^{(k)}))$$

    for this system, using $x^{(0)} = (1, 1)^T$.

    (c) Which one do you think is going to converge faster?

4. Write a computer code to implement the fixed point iterations outlined in the previous problem. Compare the total "cost to convergence" with your results for Newton's method and the chord iterations.

Solution: The MATLAB script below converges in 24 iterations to the fixed point

$$\alpha = (-0.48336323789790, 0.21344115515536)^T.$$

```
function y = fixed2(x)
    K = [2 -1; -1 2];
    b = [-1 1]';
    for k = 1:100
    k
    y = 0.5*(b - (1/9)*exp(-x) - K*x + 2*x);
    if abs(x - y) < 1.e-6
        break
    end
    x = y;
    end
```

◁ • • • ▷

7.11   APPLICATION: NUMERICAL SOLUTION OF NONLINEAR BVP'S

Exercises:

1. Set up the nonlinear system for the example (7.36)–(7.37), and verify that the result given in the text (7.38)–(7.40) is correct.

2. Apply Newton's method and the chord method, to the approximate solution of the nonlinear BVP (7.36)–(7.37). Compare the number of iterations to converge and the overall cost of convergence. Use the sequence of grids $h^{-1} = 4, 8, \ldots, 1024$.

Solution: For Newton's method with $h = 1/8$ and $u^{(0)} = 0$, we get convergence in 10 iterations to $u = (0.1267, 0.2529, 0.3787, 0.5039, 0.6287, 0.7529, 0.8767)^T$.

3. Consider the nonlinear BVP

$$-u'' + e^{-u} = 1,$$
$$u(0) = u(1) = 1.$$

Use finite difference techniques to reduce this (approximately) to a system of non-linear algebraic equations, and solve this system using several of the methods discussed in this chapter of the text. Test the program on the sequence of grids $h^{-1} = 4, 8, \ldots, 1024$ (and further, if practical on your system). Compare the cost of convergence for each method, in terms of the number of iterations and in terms of the number of operations.

Solution: The system of equations is defined by

$$-u_{k-1} + 2u_k - u_{k+1} + h^2 e^{-u_k} = h^2,$$

for $1 \leq k \leq n - 1$, with $u_0 = u_n = 1$. This leads to a nonlinear system of the form

$$F(u) = Ku + h^2\phi(u) - h^2 - g,$$

where $K = \text{tridiag}(-1, 2, -1)$, $\phi(u) = e^{-u}$, and

$$g = (1, 0, 0, \ldots, 0, 0, 1)^T.$$

The table below shows the number of iterations to convergence for several of the methods discussed in this section of the text, as a function of h. All of the methods used $u^{(0)} = 0$ as the initial guess.

h^{-1}	Newton	Chord	Fixed point
4	4	7	6
8	4	7	6
16	4	7	6
32	4	7	6

4. Now consider the nonlinear BVP

$$-u'' = \frac{u}{u + 1},$$
$$u(0) = 0, \quad u(1) = 1.$$

Repeat the kind of study required in the previous problem.

Solution: For Newton's method with $h = 1/8$ and $u^{(0)} = 0$, we get convergence in 7 iterations to $u = (0.1398, 0.2777, 0.4122, 0.5421, 0.6666, 0.7847, 0.8961)^T$.

5. Verify that (7.41) gives the correct gradient for (7.38)–(7.40).

Solution: This is a direct computation. We have that

$$[\phi'(u)]_{ij} = \frac{\partial \phi_i}{\partial u_j},$$

from which (7.41) directly follows.

CHAPTER 8

APPROXIMATE SOLUTION OF THE ALGEBRAIC EIGENVALUE PROBLEM

8.1 EIGENVALUE REVIEW

Exercises:

1. For each matrix below, find the characteristic polynomial and the eigenvalues by a hand calculation. For some of the exercises, the correct eigenvalues are given, to four decimal places, so you can check your work.

   (a)

   $$A = \begin{bmatrix} 4 & 1 & 0 \\ 1 & 4 & 1 \\ 0 & 1 & 4 \end{bmatrix}$$

   for which $\sigma(A) = \{5.4142, 4.0000, 2.5858\}$;

   Solution: $p(\lambda) = -\lambda^3 + 12\lambda^2 - 46\lambda + 56.$

   (b)

   $$A = \begin{bmatrix} 1 & 2 & 0 \\ 0 & 4 & 5 \\ 0 & 5 & 7 \end{bmatrix}$$

   for which $\sigma(A) = \{1.0000, 0.2798, 10.7202\}$;

   Solution: $p(\lambda) = -\lambda^3 + 12\lambda^2 - 14\lambda + 3.$

(c)

$$A = \begin{bmatrix} 6 & -2 & 0 \\ 2 & 6 & -2 \\ 0 & 2 & 4 \end{bmatrix}$$

for which $\sigma(A) = \{5.5698 \pm 2.6143i, 4.8603\}$;

Solution: $p(\lambda) = -\lambda^3 + 16\lambda^2 - 92\lambda + 184$.

(d)

$$A = \begin{bmatrix} 4 & 2 & 0 \\ 1 & 4 & 2 \\ 0 & 1 & 4 \end{bmatrix}$$

for which $\sigma(A) = \{2, 4, 6\}$;

Solution: $p(\lambda) = -\lambda^3 + 12\lambda^2 - 44\lambda + 48$.

(e)

$$A = \begin{bmatrix} 6 & -1 & 0 \\ 2 & 6 & -1 \\ 0 & 2 & 4 \end{bmatrix} ;$$

(f)

$$A = \begin{bmatrix} 0 & 2 & 0 \\ 2 & 7 & 1 \\ 0 & 1 & 4 \end{bmatrix} .$$

2. Prove that similar matrices have identical eigenvalues and related eigenvectors.

Solution: If $A = P^{-1}BP$, then we have

$$Ax = \lambda x \Leftrightarrow P^{-1}BPx = \lambda x \Leftrightarrow B(Px) = \lambda(Px) \Leftrightarrow By = \lambda y.$$

3. Apply Gerschgorin's Theorem to the matrices in Problem 1 and determine the intervals or disks in which the eigenvalues must lie.

Solution: For (a), $\lambda \in \{z \in \mathbf{C} \mid |z - 4| \leq 3\}$.

4. Show that the matrix

$$A = \begin{bmatrix} 3 & 2 & -1 \\ 0 & 2 & 2 \\ 0 & -2 & 7 \end{bmatrix},$$

has repeated eigenvalues and is defective, but that the matrix

$$A = \begin{bmatrix} -1 & -1 & 2 \\ -2 & 0 & 2 \\ -4 & -2 & 5 \end{bmatrix},$$

which also has repeated eigenvalues, is *not* defective.

Solution: For the first matrix, we get $\sigma(A) = \{3, 3, 6\}$, but the only eigenvector for $\lambda = 3$ is parallel to $x = (1, 0, 0)^T$; for the second matrix, we get $\sigma(A) = \{2, 1, 1\}$, and there are two independent eigenvectors for $\lambda = 1$.

5. Let A and B be similar matrices of size $n \times n$, with

$$B = \begin{bmatrix} \lambda & a^T \\ 0 & A_{22} \end{bmatrix}$$

where $\lambda \in \mathbb{R}$, $a \in \mathbb{R}^{n-1}$, and $A_{22} \in \mathbb{R}^{n-1 \times n-1}$.

(a) Prove that $\lambda \in \sigma(A)$. Hint: What is the product Be_1, where e_1 is the first standard basis vector?

Solution: Since $Be_1 = \lambda e_1$, it follows that $\lambda \in \sigma(B)$, but the similarity then implies that $\lambda \in \sigma(A)$.

(b) Prove that each eigenvalue of A_{22} is also an eigenvalue of A.

Solution: Let $\mu \in \sigma(A_{22})$, with eigenvector y. Then it is easy to show that

$$x = \begin{bmatrix} 0 \\ y \end{bmatrix}$$

is an eigenvector of B, corresponding to the same eigenvalue. The similarity of A and B then proves that $\mu \in \sigma(A)$.

6. Generalize the above; let A and B be similar matrices of size $n \times n$, with

$$B = \begin{bmatrix} D & a^T \\ 0 & A_{22} \end{bmatrix}$$

where $D \in \mathbb{R}^{p \times p}$ is diagonal, $a \in \mathbb{R}^{n-p}$, and $A_{22} \in \mathbb{R}^{n-p \times n-p}$. Prove that each diagonal element of D is an eigenvalue of A: $d_{ii} \in \sigma(A)$, $1 \leq i \leq p$, and that each eigenvalue of A_{22} is also an eigenvalue of A.

Solution: This is an obvious extension of the argument in Problem 5.

7. Let A be given by

$$A = \begin{bmatrix} 2 & 0 \\ a & A_{22} \end{bmatrix}.$$

Prove that $2 \in \sigma(A)$.

8. Let $A_1 = BC$ and $A_2 = CB$ for given B, C. Show that any non-zero eigenvalue of A_1 is also an eigenvalue of A_2, and vice-versa.

Solution: Let μ be an eigenvalue of A_1. Then we have $A_1 x = \mu x$ for an eigenvector x. Now multiply by C to get

$$CA_1 x = \mu Cx \Rightarrow CBCx = \mu Cx \Rightarrow A_2 Cx = \mu Cx \Rightarrow A_2 y = \mu y$$

for $y = Cx$. Therefore μ is an eigenvalue of A_2. A similar proof works for the other direction.

◁ • • ▷

8.2 REDUCTION TO HESSENBERG FORM

Exercises:

1. Show that in the Householder construction, if we use $c = -\|x\|_2$ then we get $Qx = -\|x\|_2 e_1$, and that this will work just as well to construct the Hessenberg matrix.

2. Compute (by hand) the Hessenberg form of the matrix

$$A = \begin{bmatrix} 6 & 1 & 1 & 1 \\ 1 & 6 & 1 & 1 \\ 1 & 1 & 6 & 1 \\ 1 & 1 & 1 & 6 \end{bmatrix}.$$

You should get

$$A_H = \begin{bmatrix} 6.0000 & 1.7321 & 0 & 0 \\ 1.7321 & 8.0000 & -0.0000 & 0 \\ 0 & -0.0000 & 5.0000 & 0 \\ 0 & 0 & 0 & 5.0000 \end{bmatrix}.$$

3. Complete the computation in Example 8.6 by finding the matrix P such that $A = PA_HP^T$.

4. Use the hess command to compute the Hessenberg form for each of H_4, H_8, and H_{10} (defined at the beginning of the Exercises for §7.2). Verify that the original matrix can be recovered from A_H.

8.3 POWER METHODS

Exercises:

1. Explain, in your own words, why it is necessary to scale the power iterations. Hint: What happens if we don't scale the iteration?

   Solution: If the iteration is not scaled, then (unless the dominant eigenvalue has absolute value equal to 1) the vector iterates will either go off to infinity or down to zero. Either way, the iteration quickly becomes useless.

2. Use an inductive argument to show that in the basic power method (Theorem 8.5) we have

$$z^{(k)} = c_k A^k z^{(0)}.$$

   What is c_k?

   Solution: We have

$$z^{(k)} = \frac{1}{\mu_k} A z^{(k-1)}$$

   so

$$z^{(k)} = \frac{1}{\mu_k \mu_{k-1}} A z^{(k-2)}$$

   and

$$z^{(k)} = \frac{1}{\mu_k \mu_{k-1} \cdots \mu_1} A z^{(0)}.$$

So

$$c_k = \frac{1}{\mu_k \mu_{k-1} \cdots \mu_1}.$$

3. Consider the iteration defined by

$$y^{(k+1)} = Az^{(k)},$$
$$z^{(k+1)} = y^{(k+1)}/\sigma_{k+1},$$

where σ_{k+1} is some scaling parameter. Assume that the iteration converges in the sense that $z^{(k)} \to z$; prove that σ_k must also converge, and that therefore (σ, z), where σ is the limit of the σ_k, must be an eigenpair of A.

Solution: We have

$$z^{(k+1)} = \frac{1}{\sigma_k} Az^{(k)}$$

so that

$$Az^{(k)} = \sigma_k z^{(k+1)};$$

therefore,

$$Az + A\left(z^{(k)} - z\right) = \sigma_k z + \sigma_k \left(z^{(k+1)} - z\right).$$

Since the two differences go to zero, it follows that σ_k converges to a limit, and we are essentially done.

4. Fill in the details of the eigenvalue error estimate in Theorem 8.5. In particular, show that there is a constant C such that

$$|\lambda_1 - \mu_k| \le C \left(\frac{\lambda_2}{\lambda_1}\right)^k.$$

5. Prove Theorem 8.7.

6. Write a program that does the basic power method. Use it to produce a plot of the largest eigenvalue of the T_n family of matrices, as a function of n, over the range $2 \le n \le 20$.

7. Write a program that does the inverse power method. Use it to produce a plot of the smallest eigenvalue of the H_n family of matrices, as a function of n, over the range $2 \le n \le 20$.

8. Write a program that does inverse power iteration on a symmetric matrix. Assume the matrix is tridiagonal, and store only the necessary non-zero elements of the matrix. Test it on the T_n and K_n families, and on your results from finding the Hessenberg forms for the H_n matrices. Produce a plot of the smallest eigenvalues of each of these matrices, as a function of n.

9. Add shifts to your inverse power iteration program. Test it on the same examples, using a selection of shifts.

10. Let λ and x be an eigenvalue and corresponding eigenvector for $A \in \mathbb{R}^{n \times n}$, and let Q be an orthogonal matrix such that $Qx = e_1$. Assume that $\|x\|_2 = 1$. Show that

$$B = QAQ^T = \begin{bmatrix} \lambda & a^T \\ 0 & A_2 \end{bmatrix}.$$

Hint: Consider each of the inner products (e_i, Be_j), using the fact that $Qx = e_1$ and that Q is symmetric.

Solution: Following the hint, we have

$$(e_i, Be_j) = (e_i, QAQ^T e_j) = (Q^T e_i, AQ^T e_j).$$

Now, $Qx = e_1$ implies that $x = Q^T e_1$ therefore the $(1, 1)$ element of B is $(e_1, Be_1) = (Q^T e_1, AQ^T e_1) = (x, Ax) = \lambda$. The rest follows with a little attention to detail.

11. An alternate deflation for symmetric matrices is based on the relationship

$$A' = A - \lambda_1 x_1 x_1^T.$$

If $A \in \mathbb{R}^{n \times n}$ is symmetric, $\lambda_1 \in \sigma(A)$ with corresponding eigenvector x_1, and if $\|x_1\|_2 = 1$, show that A' has the same eigenvalues as A, except that λ_1 has been replaced by zero. In other words, if

$$\sigma(A) = \{\lambda_1, \lambda_2, \ldots, \lambda_n\},$$

then

$$\sigma(A') = \{0, \lambda_2, \ldots, \lambda_n\}.$$

Solution: Since A is symmetric, the eigenvectors can be assumed orthonormal, thus

$$A' x_k = A x_k - \lambda_1 x_1 x_1^T x_k = A x_k - \lambda \delta_{1k} = \lambda_k - \lambda \delta_{1k}.$$

which is sufficient.

12. Show how to implement the deflation from the previous problem (known as Hotelling's deflation) without forming the product $x_1 x_1^T$. Hint: What is $A'z$ for any vector z? Can we write this in terms of the dot product $x_1^T z$? (According to Wilkinson [14], this deflation is prone to excessive rounding error and so is not of practical value.)

13. What is the operation count for one iteration of the basic power method applied to a symmetric matrix? Compute two values, one for the matrix in Hessenberg form, one for the full matrix.

Solution: The iteration is

(a) Compute $y^{(k)} = Az^{(k-1)}$;

(b) $\mu_k = y_i^{(k)}$, where $\|y^{(k)}\|_\infty = |y_i^{(k)}|$;

(c) Set $z^{(k)} = y^{(k)}/\mu_k$.

This amounts to a single matrix-vector multiplication, followed by a scalar-vector multiplication. (We will ignore the cost of determining the maximum.) So the cost of each iteration is

$$C = n^2 + n$$

multiplications. If the matrix is in Hessenberg form, then the matrix-vector multiplication only takes $3n$ operations, for a much cheaper iteration.

14. What is the cost of the basic power method for a symmetric matrix if it runs for N iterations. Again, compute two values depending on whether or not a Hessenberg reduction is done. Is it always the case that a reduction to Hessenberg form is cost-effective?

15. Repeat Problems 13 and 14 for the inverse power method.

16. Consider the family of matrices T defined by

$$
t_{ij} = \begin{cases}
a, & i - j = 1; \\
b, & i = j; \\
c, & i - j = -1; \\
0, & \text{otherwise.}
\end{cases}
$$

Thus T is tridiagonal, with constant diagonal elements a, b, and c. It is commonplace to abbreviate such a matrix definition by saying

$$T = \operatorname{tridiag}(a, b, c).$$

Write a computer program that uses the power method and inverse power method to find the largest and smallest (in absolute value) eigenvalues of such a matrix, for any user-specified values of a, b, and c. Test the program on the following examples; note that the exact values are given (to four places):

(a) $a = c = 4$, $b = 1$, $n = 6$, $\lambda_1 = -6.2708$, $\lambda_6 = -0.7802$;

(b) $a = b = c = 6$, $n = 7$, $\lambda_1 = 17.0866$, $\lambda_6 = 1.4078$;

(c) $a = 2$, $b = 3$, $c = 4$, $n = 5$, $\lambda_1 = 7.8990$, $\lambda_6 = 0.1716$;

(d) $a = c = 1$, $b = 10$, $n = 9$, $\lambda_1 = 11.9021$, $\lambda_6 = 8.0979$;

(e) $a = c = 1$, $b = 4$, $n = 8$, $\lambda_1 = 5.8794$, $\lambda_6 = 2.1206$.

17. Modify the program from the previous problem to do the inverse power method with shifts. Test it on the same examples by finding the eigenvalue closest to $\mu = 1$.

8.4 BISECTION AND INERTIA TO COMPUTE EIGENVALUES OF SYMMETRIC MATRICES

Exercises:

1. For each of the following matrices, use the hess command to reduce it to tridiagonal form, then construct an LDL^T factorization to find the inertia of the matrix.

$$
A_1 = \begin{bmatrix}
5 & 1 & 0 & 0 & 0 & 1 \\
1 & 4 & 1 & 0 & 0 & 0 \\
0 & 1 & 4 & 1 & 0 & 0 \\
0 & 0 & 1 & 4 & 1 & 0 \\
0 & 0 & 0 & 1 & 4 & 1 \\
1 & 0 & 0 & 0 & 1 & 4
\end{bmatrix};
\quad
A_2 = \begin{bmatrix}
-2 & -10 & 4 & -7 & 1 \\
-10 & -5 & -1 & -3 & -3 \\
4 & -1 & 2 & 0 & -8 \\
-7 & -3 & 0 & 0 & -10 \\
1 & -3 & -8 & -10 & 5
\end{bmatrix};
$$

$$A_3 = \begin{bmatrix} 2 & 1 & 1 & 1 & 1 \\ 1 & 0 & 1 & 0 & 1 \\ 1 & 1 & 1 & 2 & 1 \\ 1 & 0 & 2 & 2 & 1 \\ 1 & 1 & 1 & 1 & 2 \end{bmatrix} ; \quad A_4 = \begin{bmatrix} 1 & 2 & 3 & 2 & 2 \\ 2 & 4 & 1 & 2 & 1 \\ 3 & 1 & 2 & 2 & 3 \\ 2 & 2 & 2 & 1 & 1 \\ 2 & 1 & 3 & 1 & 4 \end{bmatrix} .$$

Solution: For A_1, we get

$$A_{1,H} = \begin{bmatrix} 5.1250 & 0.5995 & 0 & 0 & 0 & 0 \\ 0.5995 & 3.3533 & 0.2750 & 0 & 0 & 0 \\ 0 & 0.2750 & 3.9217 & 1.3565 & 0 & 0 \\ 0 & 0 & 1.3565 & 4.1000 & 1.1180 & 0 \\ 0 & 0 & 0 & 1.1180 & 4.5000 & -1.4142 \\ 0 & 0 & 0 & 0 & -1.4142 & 4.0000 \end{bmatrix} ,$$

and $A_{1,H} = L_1 D_1 L_1^T$, for

$$D_1 = \mathrm{diag}(5.1250, 3.2831, 3.8987, 3.6280, 4.1555, 3.5187),$$

thus all six eigenvalues of A_1 are positive. (Gerschgorin's Theorem tells us the same thing.)

For A_3, we get

$$A_{3,H} = \begin{bmatrix} 0.5000 & -0.8660 & 0 & 0 & 0 \\ -0.8660 & 0.1667 & -0.9428 & 0 & 0 \\ 0 & -0.9428 & 0.0833 & 1.2990 & 0 \\ 0 & 0 & 1.2990 & 4.2500 & -2.0000 \\ 0 & 0 & 0 & -2.0000 & 2.0000 \end{bmatrix} ,$$

and $A_{3,H} = L_3 D_3 L_3^T$, for

$$D_3 = \mathrm{diag}(0.5000, -1.3333, 0.7500, 2.0000, -0.0000).$$

If we ask for more digits, we see that the fifth element of the diagonal is actually very close to zero, but negative. We thus have one negative eigenvalue, three positive eigenvalues, and one that we frankly are not sure of, but suspect it may be a second negative eigenvalue.

2. For each of the matrices in the previous problem, carry out a "manual" iteration to isolate each eigenvalue in its own bin, similar to what we did in the first examples of this section. That is, choose a shift, factor the shifted matrix into LDL^T form, update the bins, and continue until each eigenvalue is isolated in its own bin. (Please don't "cheat" by running eig on the matrix beforehand to see where to put the shifts.)

Solution: For A_1, the "Gerschgorin interval" is $[2, 7]$. Trying various shifts we get the following results:

$$\sigma = 4.5; \quad D_\sigma = (0.6250, -1.7217, -0.5343, 3.0435, -0.4107, 4.3696),$$
$$\sigma = 3.25; \quad D_\sigma = (1.8750, -0.0884, 1.5270, -0.3549, 4.7717, 0.3309),$$
$$\sigma = 2.625; \quad D_\sigma = (2.5000, 0.5845, 1.1674, -0.1012, 14.2286, 1.2344).$$

The third shift tells us there is exactly one eigenvalue $\lambda_1 \in [2, 2.625]$; the second shift then tell us there is exactly one eigenvalue $\lambda_2 \in [2.625, 3.25]$, and the first shift then tells us there is exactly one eigenvalue $\lambda_2 \in [3.25, 4.5]$. We thus have

$$\text{Bin}_1 = (2 \leq \lambda_1 \leq 2.625), \qquad \text{Bin}_2 = (2.625 \leq \lambda_2 \leq 3.25),$$
$$\text{Bin}_3 = (3.25 \leq \lambda_3 \leq 4.5).$$

Now try a few more shifts:

$$\sigma = 5.75; \quad D_\sigma = (-0.6250, -1.8217, -1.7868, -0.6202, 0.7655, -4.3627),$$
$$\sigma = 5.125; \quad D_\sigma = \text{Oops!}$$

Using $\sigma = 5.125$ gives us a nearly singular matrix, so we will perturb the shift a bit, but before we try that, we note that $\sigma = 5.75$ told us there is exactly one eigenvalue $\lambda_6 > 5.75$, thus we have $\text{Bin}_6 = (5.75 \leq \lambda_6 \leq 7)$, and therefore there are exactly two eigenvalues satisfying

$$4.5 \leq \lambda_5, \lambda_6 \leq 5.75.$$

So, we try $\sigma = 5.2$ and get

$$\sigma = 5.2; \quad D_\sigma = (-0.0750, 2.9449, -1.3039, 0.3111, -4.7179, -0.7761),$$

which tell us there is one eigenvalue in $[4.5, 5.2]$ and one in $[5.2, 5.75]$. Thus, our complete set of bins is

$$\text{Bin}_1 = (2 \leq \lambda_1 \leq 2.625), \qquad \text{Bin}_2 = (2.625 \leq \lambda_2 \leq 3.25),$$
$$\text{Bin}_3 = (3.25 \leq \lambda_3 \leq 4.5), \qquad \text{Bin}_4 = (4.5 \leq \lambda_4 \leq 5.2),$$
$$\text{Bin}_5 = (5.2 \leq \lambda_5 \leq 5.75), \qquad \text{Bin}_6 = (5.75 \leq \lambda_6 \leq 7).$$

The exact eigenvalues are

$$\lambda = 2.1088, \ 3.0000, \ 3.2954, \ 5.0000, \ 5.3174, \ 6.2784,$$

which confirms that the bins are correct.

For A_3, the "Gerschgorin interval" is $[-4, 6]$. We will start with the shifts

$$\sigma = 1; \quad D_\sigma = (-0.5000, 0.6667, -2.2500, 4.0000, 0.0000),$$
$$\sigma = 3.5; \quad D_\sigma = (-3.0000, -3.0833, -3.1284, 1.2894, -4.6022),$$
$$\sigma = 4.75; \quad D_\sigma = (-4.2500, -4.4069, -4.4650, -0.1221, 30.0215).$$

The last shift tell us there is exactly one eigenvalue

$$4.75 \leq \lambda_5 \leq 6,$$

but the second shift tells us nothing new. Worse, the first shift only muddies the water because we can't really be sure if that zero is a zero or small positive/negative. (Something not emphasized enough in the text is that the inertia approach does not like results near zero, because they often do not yield solid information.) But, we soldier on:

$$\sigma = -1.5; \quad D_\sigma = (2.0000, 1.2917, 0.8952, 3.8649, 2.4650).$$

This illustrates how Gerschgorin's Theorem can give us too large a region where the eigenvalues might be. Since the inertia here is completely positive, we know that all five eigenvalues are ≥ -1.5, and four of them are in $[-1.5, 4.75]$. Let's work on just that shorter interval now, trying the following:

$$\sigma = 1.625; \quad D_\sigma = (-1.1250, -0.7917, -0.4189, 6.6538, -0.2262),$$

This might be called frustrating—we have isolated no additional eigenvalues—but we are narrowing the intervals where the eigenvalues live. We already know that

$$4.75 \leq \lambda_5 \leq 6,$$

but we also now know that the other four eigenvalues all satisfy

$$-4 \leq \lambda_1, \lambda_2, \lambda_3, \lambda_4 \leq 1.625.$$

So we now try

$$\sigma = -1.1875; \quad D_\sigma = (1.6875, 0.9097, 0.2937, -0.3075, 16.1960).$$

This tells us there is only one eigenvalue less than -1.1875, thus we have

$$-4 \leq \lambda_1 \leq -1.1875$$

and

$$-1.1875 \leq \lambda_2, \lambda_3, \lambda_4 \leq 1.625.$$

Let's try the midpoint of this interval

$$\sigma = 0.21875; \quad D_\sigma = (0.2813, -2.7187, 0.1915, -4.7793, 2.6182).$$

This tell us that two eigenvalues are less than 0.21875, but we already know that one is ≤ -1.1875, so we have

$$-1.1875 \leq \lambda_2 \leq 0.21875,$$

and our bins (so far) look like this:

$$\text{Bin}_1 = (-4 \leq \lambda_1 \leq -1.1875), \qquad \text{Bin}_2 = (-1.1875 \leq \lambda_2 \leq 0.21875),$$
$$\text{Bin}_5 = (4.75 \leq \lambda_5 \leq 6),$$

with

$$0.21875 \leq \lambda_3, \lambda_4 \leq 4.75.$$

We are pleased that "the eigenvalue that might be zero" has already been isolated (this is λ_2). To try to complete the bins, we start with the midpoint of the remaining interval that contains two eigenvalues:

$$\sigma = 2.484375; \quad D_\sigma = (-1.9844, -1.9398, -1.9428, 2.6342, -2.0029).$$

The inertia tells there are four eigenvalues ≤ 2.484375, and only one larger than that; this doesn't help us any, so we do another bisection

$$\sigma = 1.3515625; \quad D_\sigma = (-0.8516, -0.3042, 1.6542, 1.8783, -1.4811),$$

and now we have success—there are three eigenvalues ≤ 1.3515625, so we now know that

$$0.21875 \leq \lambda_3 \leq 1.3515625$$

and

$$1.3515625 \leq \lambda_4 \leq 4.75,$$

thus our final set of bins is

$$\text{Bin}_1 = (-4 \leq \lambda_1 \leq -1.1875), \qquad \text{Bin}_2 = (-1.1875 \leq \lambda_2 \leq 0.21875),$$
$$\text{Bin}_3 = (0.21875 \leq \lambda_3 \leq 1.3515625), \quad \text{Bin}_4 = (1.3515625 \leq \lambda_4 \leq 4.75),$$
$$\text{Bin}_5 = (4.75 \leq \lambda_5 \leq 6).$$

The exact eigenvalues are

$$\lambda = (-1.2332, 0.0000, 1.0000, 1.5770, 5.6562)$$

(and that is an exact zero eigenvalue—A_3 is singular), and we can see that each bin contains a single eigenvalue.

3. Write a script that uses the bisection/inertia method to find all eigenvalues/eigenvectors of a symmetric tridiagonal matrix. A good outline of the code might be as follows:

   (a) A function `Bins0` which computes the initial set of bins over the Gerschgorin interval.

   (b) A function `BinsF` which takes the original bins and refines them (using bisection) to get a set of bins that each contain a single eigenvalue.

   (c) A function `Solve` which uses inverse iteration (Algorithm 8.3) to compute the eigenpair corresponding to a given bin containing one eigenvalue.

   Test your script on a set of examples and compare its performance to the native routines in the computing environment you use.

Solution: This is a very challenging code to write, mostly because the process of isolating the eigenvalues (constructing the bins) is so very involved. In the text, we used a "multi-section" approach; here in the Solutions Manual we stuck with true bisection. Based on our results, the multisection idea appears to be better. Good luck!

8.5 AN OVERVIEW OF THE QR ITERATION

Exercises:

1. Show that the product of two orthogonal matrices, Q_1 and Q_2, is also orthogonal.

   Solution:
   $$(Q_1 Q_2)^T (Q_1 Q_2) = Q_2^T Q_1^T Q_1 Q_2 = Q_2^T (I) Q_2 = I.$$

2. Let $A \in \mathbb{R}^{n \times n}$ have the partitioned form

$$A \begin{bmatrix} A_{11} & a \\ 0 & \lambda \end{bmatrix}$$

where $A_{11} \in \mathbb{R}^{(n-1) \times (n-1)}$ and $\lambda \in \mathbb{R}$. Show that λ must be an eigenvalue of A. How are the eigenvalues of A and A_{11} related?

Solution: For A as given, let $e_n = (0, 0, 0, \ldots, 0, 1)^T$. Then

$$A e_n = \lambda e_n$$

so λ must be an eigenvalue of A. Now let z be an arbitrary eigenvector of A_{11} and form $x = (z, 0)^T$. Then

$$Ax = \begin{pmatrix} A_{11} z \\ 0 \end{pmatrix}$$

so all the eigenvalues of A_{11} are also eigenvalues of A.

3. Assume that the iteration in Algorithm 8.5 converges, in the sense that $R_k \to R_\infty$ which is upper triangular, and $Q_k \to Q_\infty$ which is orthogonal.

(a) Prove that the matrices A_k, defined in (8.5), must also converge to R_∞.

Solution: We have

$$A_k = Q_k^T Q_{k+1} Q_{k+1}^T A Q_k = Q_k^T Q_{k+1} R_{k+1}.$$

The rest of the proof should be easy.

(b) Prove then that the eigenvalues of A can be recovered from the diagonal elements of A_k, in the limit as $k \to \infty$.

Solution: Since $A_k \to R_\infty$, and A_k is similar to A, it should be easy to construct a formal argument.

4. Show that the matrices in the shifted QR iteration all have the same eigenvalues.

Solution: We have

$$Q_k R_k = A_{k-1} - \mu_{k-1} I$$

so that

$$Q_k R_k Q_k = A_{k-1} Q_k - \mu_{k-1} Q_k;$$

thus,

$$Q_k (A_k - \mu_{k-1} I) = A_{k-1} Q_k - \mu_{k-1} Q_k,$$

or,

$$Q_k A_k = A_{k-1} Q_k$$

and we are done.

5. Consider the matrix

$$A = \begin{bmatrix} 0 & 1 \\ 1 & 0 \end{bmatrix}$$

(a) Find the exact eigenvalues of A;

(b) Show that

$$Q = \begin{bmatrix} 0 & 1 \\ 1 & 0 \end{bmatrix}, \quad R = \begin{bmatrix} 1 & 0 \\ 0 & 1 \end{bmatrix}$$

is a valid QR decomposition of A;

(c) Use this to show that the shifted QR iteration for A, using the Rayleigh shift, will not converge.

Solution: (a) The eigenvalues are easily shown to be $\lambda_1 = 1$, $\lambda_2 = -1$. (b) $QR = A$, therefore we are done. (c) The Rayleigh shift is $a_{22} = 0$, therefore the first step of QR using this shift simply reproduces the original matrix: $A_1 = RQ = A$. Hence, the iteration fails to converge; in fact, it fails to do much of anything!

6. The QR factorization is usually carried out for the QR iteration by means of *Givens transformations*, defined in part (a), below. In this exercise we will introduce the basic ideas of this kind of matrix operation.

   (a) Show that the matrix

   $$G(\theta) = \begin{bmatrix} \cos\theta & -\sin\theta \\ \sin\theta & \cos\theta \end{bmatrix}$$

   is an orthogonal matrix for any choice of θ.

   (b) Show that we can always choose θ so that

   $$G(\theta)A = \begin{bmatrix} \cos\theta & -\sin\theta \\ \sin\theta & \cos\theta \end{bmatrix} \begin{bmatrix} a_{11} & a_{12} \\ a_{21} & a_{22} \end{bmatrix} = \begin{bmatrix} r_{11} & r_{12} \\ 0 & r_{22} \end{bmatrix} = R$$

   (c) Use a Givens transformation to perform one step of the basic QR iteration for the matrix

   $$\begin{bmatrix} 4 & 1 \\ 1 & 4 \end{bmatrix}.$$

7. (a) If $G(\theta)$ is a Givens transformation, show that

   $$\hat{G}(\theta) = \begin{bmatrix} 1 & 0 \\ 0 & G(\theta) \end{bmatrix}$$

   is orthogonal.

   (b) Show how to use a sequence of *two* Givens transformations to do a single QR step for the matrix

   $$\begin{bmatrix} 4 & 1 & 0 \\ 1 & 4 & 1 \\ 0 & 1 & 4 \end{bmatrix}.$$

8. Write a MATLAB code that automates the kind of computation we did in Examples 8.17 and 8.18: Take an input matrix (square), and perform "naive" QR by executing the commands:

```
[q, r] = qr(A);
A = r * q;
```

Stop when A is triangular. Use no shifts. Test your code by forming a random symmetric matrix[1] and using your routine to find its eigenvalues. Remember to reduce the matrix to Hessenberg form! Use the appropriate timing routines to measure how long your codes takes to do an decent-sized example, and compare this to what MATLAB's `eig` command costs.

Solution: Here is a MATLAB code that does this on a 6×6 (symmetric) example:

```
B = rand(6,6);
A = (B + B')/2
tic
AH = hess(A);
Asave = AH;
offdiag = 1;
k = 0;
while offdiag > 1.e-16
k = k + 1;
[Q,R] = qr(AH);
AH = R*Q;
AH;
v = diag(AH,-1);
offdiag = norm(v)
end
diag(AH)
TA = toc
tic
lambda = eig(A)
TM = toc
R = TA/TM
```

When the author ran this, `eig` was consistently 2–3 orders of magnitude faster than this routine. Obviously, a good shifting strategy (and deflation) helps a lot!

9. Repeat the above, this time using the Rayleigh shift. (You should save the random matrix from the previous problem so a timing comparison is meaningful.)

10. Repeat the above, this time using the Wilkinson shift. (Again, use the same matrix and do a timing comparison.)

◁ ● ● ● ▷

8.6 APPLICATION: ROOTS OF POLYNOMIALS, II

Exercises:

1. For the polynomial $p(x) = x^3 - 2x^2 + 5x + 1$, construct the companion matrix, then use a cofactor expansion to confirm that the characteristic polynomial is, indeed, $p(\lambda)$.

---

[1]Recall that $S = A + A^T$ is always symmetric.

2. For each polynomial in Problem 2 of §3.10, construct the companion matrix and use MATLAB's `eig` command to find the roots. Use the appropriate timing commands to compare the speed with the Durand–Kerner method from §3.10.

Solution: For (b) $(p(x) = x^6 - x^5 - 1)$, the companion matrix is

$$C = \begin{bmatrix} 0 & 0 & 0 & 0 & 0 & 1 \\ 1 & 0 & 0 & 0 & 0 & 0 \\ 0 & 1 & 0 & 0 & 0 & 0 \\ 0 & 0 & 1 & 0 & 0 & 0 \\ 0 & 0 & 0 & 1 & 0 & 0 \\ 0 & 0 & 0 & 0 & 1 & 1 \end{bmatrix}.$$

The author's Durand–Kerner routine from Chapter 3 found the roots in about 8.71165×10^{-5} seconds; using `eig` on the companion matrix took about 9.28236×10^{-6} seconds (much faster), whereas using the `roots` command directly on the polynomial took 2.017166×10^{-5} seconds. All three methods found the roots in different orders, which made a comparison a tad tricky, but they all found the same roots. These timing estimates ares based on an average of 5000 `tic-toc` values for each method, run with no other processes active.

3. Consider the polynomial $p(x) = x^4 - 10x^3 + 35x^2 - 50x + 24$ and form its companion matrix. Using MATLAB's `qr` command, do several iterations of the unshifted QR iteration. Are the roots being isolated? Is the structure of the companion matrix being maintained? Comment.

◁ ● ● ▷

8.7 APPLICATION: COMPUTATION OF GAUSSIAN QUADRATURE RULES

Exercises:

1. Following the general scheme we used in the examples here, construct the 4-point Gauss–Laguerre rule. Check your work by using your weights and abscissas to compute the following integrals (you should get the exact values):

(a)
$$\int_0^\infty x^2 e^{-x}\,dx = 2;$$

(b)
$$\int_0^\infty x^7 e^{-x}\,dx = 5040.$$

Solution: The "unsymmetrized" matrix is

$$K_4 = \begin{bmatrix} 1 & -1 & 0 & 0 \\ -1 & 3 & -2 & 0 \\ 0 & -2 & 5 & -3 \\ 0 & 0 & -3 & 7 \end{bmatrix},$$

which has eigenvalues

$$\lambda_i = (0.322547689619392, 1.745761101158346, 4.536620296921128, 9.395070912301129),$$

which are the correct abscissae values. We now wake up from our groggy state and realize that K_4 is, in fact symmetric, so we don't have to do any work to symmetrize it. The orthonormal eigenvectors are

$$v_1 = (-0.776629966163, -0.526129764888, -0.316028694198, -0.141983205349)^T,$$
$$v_2 = (0.597845040489, -0.445849575717, -0.578523460693, -0.330318132748)^T,$$
$$v_3 = (-0.197200173720, 0.697422136933, -0.437236418707, -0.532483585247)^T,$$
$$v_4 = (-0.023222719599, 0.194956377808, -0.611768570644, 0.766284498094)^T,$$

and the weight factor is

$$\mu = \int_0^\infty e^{-x} dx = 1,$$

so the weights are

$$w_1 = (-0.776629966162543)^2 = 0.603154104341633,$$
$$w_2 = (0.597845040489423)^2 = 0.357418692437800,$$
$$w_3 = (-0.197200173719511)^2 = 0.038887908515005,$$
$$w_4 = (-0.023222719598732)^2 = 0.000539294705561.$$

We can test the accuracy of this by using our 4-point rule to compute

$$I(p) = \int_0^\infty x^p e^{-x} dx = p!;$$

according to the theory developed in §5.6, the 4-point rule should be *exact* for any $p \le 2 \times 4 - 1 = 7$, and it is a simple computation to confirm that this is the case for these values of x_i and w_i.

2. Repeat the above for the Gauss–Chebyshev rule; this time check your values by computing the following two integrals (again, you should get the exact answers):

(a)
$$\int_{-1}^1 \frac{x^2}{\sqrt{1-x^2}} dx = \frac{\pi}{2};$$

(b)
$$\int_{-1}^1 \frac{x^4}{\sqrt{1-x^2}} dx = \frac{3\pi}{8}.$$

Solution: There is a bit of a hidden "gotcha" thing with the Chebyshev rule. If we work too quickly and carelessly, we will get

$$C_4 = \begin{bmatrix} 0 & 1/2 & 0 & 0 \\ 1/2 & 0 & 1/2 & 0 \\ 0 & 1/2 & 0 & 1/2 \\ 0 & 0 & 1/2 & 0 \end{bmatrix},$$

as our matrix, and this has eigenvalues

$$\lambda = (-0.8090, -0.3090, 0.3090, 0.8090)^T.$$

But we know that the Gauss points are supposed to be the zeros of the polynomial $T_4(x)$, and we know these from Chapter 4:

$$\xi = (-0.9239, -0.3827, 0.3827, 0.9239)^T \neq (-0.8090, -0.3090, 0.3090, 0.8090)^T.$$

What is going on here?

The first line in the matrix represents the relationship between $T_1(x)$ and $T_0(x)$, which is $T_1(x) = xT_0(x)$. Thus, that $1/2$ in the $(1, 2)$ position needs to be a 1:

$$K_4 = \begin{bmatrix} 0 & 1 & 0 & 0 \\ 1/2 & 0 & 1/2 & 0 \\ 0 & 1/2 & 0 & 1/2 \\ 0 & 0 & 1/2 & 0 \end{bmatrix}.$$

This has the correct eigenvalues, and yields the correct weights (after symmetrizing the matrix), which are all equal to $\frac{\pi}{4}$. The specific integral values are easily confirmed.

3. Extend the work here to produce a script which will ask for the number of points, and return the Gaussian quadrature nodes and weights, for each of the families we studied in this section. Verify that your scripts work by confirming that they produce the correct nodes and weights, and then by testing each of them on several examples, including the following:

(a) Gauss–Legendre: $I = \int_{-1}^{1} xe^x dx = \frac{2}{e} = 7.357588823428847 \times 10^{-1}$;

(b) Gauss–Hermite: $I = \int_{-\infty}^{\infty} xe^x e^{-x^2} dx = 1.137937897234373$;

(c) Gauss–Laguerre: $I = \int_{0}^{\infty} \sin \pi x e^{-x} dx = \frac{\pi}{\pi^2+1} = 0.2890254822222362$;

(d) Gauss–Chebyshev: $I = \int_{-1}^{1} \exp(x)(1 - x^2)^{-1/2} dx = 3.977463260506311.$

4. The n-point Gauss–Legendre rule is exact when integrating a polynomial of degree $\leq 2n - 1$. In §5.6, we showed that

$$w_i^{(n)} = \int_{-1}^{1} L_i^{(n)}(x) dx,$$

for

$$L_i^{(n)}(x) = \prod_{\substack{k=1 \\ k \neq i}}^{n} \frac{x - x_k^{(n)}}{x_i^{(n)} - x_k^{(n)}}.$$

Note that, if we want to construct an n-point rule, the integral for w involves a polynomial of degree $= n - 1 < 2n - 1$. Show how we can use this to construct the 8-point rule, given the 4-point rule plus the 8-point nodes, and then the 16-point rule from the 8-point rule plus the 16-point nodes, etc. Comment on the merits and flaws of this approach, compared to what we did in this section. (The point here is to show that we can still construct the Gauss–Legendre rules without needing to symmetrize the eigenproblem.)

Solution: *Note: Because the text went to press before the Solutions Manual was written, some issues in the phrasing of this Exercise were overlooked. A full correction/explanation will be posted on the text website.* The issue is to get the $n = 2$ weights, because if all we have is the $n = 1$ case, well, $N = 2n - 1 = 1$ does not gain us anything. If we solve the $n = 2$ eigenproblem to get the Gauss points

$$x_1^{(2)} = -0.57735\ldots = -\frac{1}{\sqrt{3}}, \quad x_2^{(2)} = 0.57735\ldots = \frac{1}{\sqrt{3}},$$

then the integrals for the weights are

$$w_i^{(n)} = \int_{-1}^{1} L_i^{(n)}(x)\,dx,$$

where

$$L_i^{(n)}(x) = \prod_{\substack{k = 1 \\ k \neq i}}^{n} \frac{x - x_k^{(n)}}{x_i^{(n)} - x_k^{(n)}},$$

so

$$w_1^{(2)} = \int_{-1}^{1} L_1^{(2)}(x)\,dx = w_2^{(2)} = \int_{-1}^{1} L_2^{(2)}(x)\,dx = 1.$$

Now we can begin some serious "bootstrapping." With the $n = 2$ rule constructed, we can integrate any polynomial of degree $\leq 2 \times 2 - 1 = 3$ exactly. If we solve the $n = 3$ eigenproblem to get the Gauss points

$$x_1^{(3)} = -0.774597\ldots = -\sqrt{\frac{3}{5}}, \quad x_2^{(3)} = 0, \quad x_3^{(3)} = 0.774597\ldots = \sqrt{\frac{3}{5}},$$

this means we can integrate any polynomial of degree $\leq 2 \times 3 - 1 = 5$. Therefore, we can get the exact weights for any rule up to $n = 6$, and we can continue onward. Symmetrizing the matrix is computationally simpler, but this process will work. (But it is only practical for the Gauss–Legendre scheme; the presence of the weight function in the other rules makes things awkward, at best.)

5. Write a script that implements the construction outlined in the previous exercise. Start with the $n = 1$ rule (which is the midpoint rule) and construct, in turn, the complete Gauss–Legendre rules for the $n = 2, 4, 8, 16$ cases. Verify that your script works by confirming that you compute the correct nodes and weights, and by testing it on a reasonable set of examples.

6. Because the presentation here was admittedly incomplete and not entirely rigorous, there are a handful of "unresolved" issues in the presentation. Make a list of each such issue, and outline how to resolve the uncertainties.

◁ • • ▷

CHAPTER 9

A SURVEY OF NUMERICAL METHODS
FOR PARTIAL DIFFERENTIAL EQUATIONS

9.1 DIFFERENCE METHODS FOR THE DIFFUSION EQUATION

Exercises:

1. Write a program to use the explicit method to solve the diffusion equation

$$
\begin{aligned}
u_t &= u_{xx}, \quad t > 0, 0 < x < 0.1; \\
u(0, t) &= 0; \\
u(1, t) &= 0; \\
u(x, 0) &= \sin \pi x;
\end{aligned}
$$

which has exact solution

$$u(x, t) = e^{-\pi^2 t} \sin \pi x.$$

(The student should check that this is indeed the exact solution.) Use $h^{-1} = 4, 8, \ldots$, and take Δt as large as possible for stability. Confirm that the approximate solution is as accurate as the theory predicts. Compute out to $t = 1$.

Solution: The following MATLAB script will solve the problem. The author makes no claims of being the most sophisticated MATLAB programmer.

```
clear
n = input('number of points? ')
```

Solutions Manual to Accompany An Introduction to Numerical Methods and Analysis, Third Edition.
James F. Epperson.
© 2021 John Wiley & Sons, Inc. Published 2021 by John Wiley & Sons, Inc.

```
h = 1/n;
dt = 0.5*h^2;
%
% Index shift since Matlab doesn't allow zero subscripts
%
np = n+1;
x = h*[0:n];
uo = sin(pi*x);
nt = floor(0.1/dt);
disp(nt)
disp('steps')
r = dt/(h*h);
pause
for k=1:nt
t = k*dt;
u(1) = 0;
        ue(1) = 0;
for j=2:n
u(j) = uo(j) + r*(uo(j-1) - 2*uo(j) + uo(j+1));
ue(j) = exp(-pi*pi*t)*sin(pi*(j-1)*h);
error(j) = ue(j) - u(j);
end
u(np) = 0;
        ue(np) = 0;
uo = u;
errmax = norm(error,inf);
disp([t,errmax])
if k == 4
figure(1)
plot(x,u)
figure(2)
plot(x,ue)
pause
end
if k == nt
figure(3)
plot(x,u)
figure(4)
plot(x,ue)
pause
end
end
```

2. Repeat the above problem, but this time take Δt 10% too large to satisfy the stability condition, and attempt to compute solutions out to $t = 1$. Comment on what happens.

   Solution: With Δt too large, the solution will quickly blow up.

3. Apply Crank-Nicolson to the same PDE as in Problem 1. For each value of h, adjust the choice of Δt to obtain comparable accuracy in the pointwise norm to what was achieved above. Comment on your results. Try to estimate the number of operations needed for each computation.

4. Write a program to use the explicit method to solve the diffusion equation

$$\begin{aligned} u_t &= u_{xx}, \quad t > 0, 0 < x < 1; \\ u(0,t) &= 0; \\ u(1,t) &= e^{-\pi^2 t/4}; \\ u(x,0) &= \sin \pi x/2; \end{aligned}$$

which has exact solution

$$u(x,t) = e^{-\pi^2 t/4} \sin \pi x/2.$$

(The student should check that this is indeed the exact solution.) Use $h^{-1} = 4, 8, \dots$, and take Δt as large as possible for stability. Confirm that the approximate solution is as accurate as the theory predicts. Compute out to $t = 1$.

Solution: The previous MATLAB script, appropriately modified to reflect the changes in boundary and initial conditions, should work.

5. Modify the three algorithms for the diffusion equation to handle non-homogeneous boundary conditions, i.e., to handle problems of the form:

$$\begin{aligned} u_t &= u_{xx} + f(x,t), \quad t > 0, 0 < x < 1; \\ u(0,t) &= g_0(t); \\ u(1,t) &= g_1(t); \\ u(x,0) &= u_0(x); \end{aligned}$$

where $g_0(t)$ and $g_1(t)$ are not identically zero. Take

$$f(x,t) = -2e^{x-t}; \quad g_0(t) = e^{-t}; \quad g_1(t) = e^{1-t}; \quad u_0(x) = e^x,$$

for which the exact solution is $u(x,t) = e^{x-t}$ (confirm this).

Solution: Using the explicit method, $h = 1/16$, and $\Delta t = 90\%$ of the maximum allowed for stability, the author got the graph in Fig. 9.1 for the maximum error at each time step (the horizontal axis is the number of time steps).

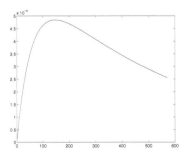

Figure 9.1   Maximum error as a function of time step number for Problem 5.

6. Recall the definition of the matrix 2-norm:

$$\|A\|_2 = \max_{u \neq 0} \frac{\|Au\|_2}{\|u\|_2},$$

where $\| \cdot \|_2$ is the usual vector 2-norm. If D is an $n \times n$ diagonal matrix, use this definition to show that

$$\|D\|_2 = \max_{1 \leq k \leq n} |d_{ii}|,$$

where d_{ii} are the diagonal elements of D.

7. Let Q be an arbitrary $n \times n$ orthogonal matrix; show that $\|Q\|_2 = 1$.

8. Compute the number of operations needed to compute out to $t = T$ using the explicit method, as a function of n, the number of points in the spatial grid. Assume that the time step is chosen to be as large as possible to satisfy the stability condition.

   Solution: Taking the time step as large as possible means

   $$\Delta t = h^2/2a,$$

   where a is the diffusion coefficient. So, to reach $t = T$ requires

   $$N_T = T/\Delta t = 2Ta/h^2 = 2Tan^2$$

   time steps. Each time step involves $3n$ multiplications, so the total cost is

   $$C = 6Tan^3.$$

9. Compute the number of operations needed to compute out to $t = T$ using Crank-Nicolson, taking $\Delta t = ch$ for some constant $c > 0$, again as a function of n, the number of points in the spatial grid.

   Solution: This time the number of time steps is

   $$N_T = T/(ch) = Tn/c.$$

   The cost of each step is the cost of forming, factoring, and solving the tridiagonal linear system, so the total cost is

   $$C = (6n)(Tn/c) = (6/c)Tn^2$$

10. Consider the *nonlinear* equation

    $$u_t = u_{xx} + Vuu_x.$$

    Take $u_0(x) = \sin \pi x$ and homogeneous boundary data, and solve this using the explicit method, taking Δt to be 90% of the maximum value allowed for stability. Compute out to $t = 0.2$, and plot your solution, for $V = 1, 5, 10$.

    Solution: The following MATLAB script should do the job:

```
n = input(' Number of points? ');
N = n+1;
Nm = N - 1;
```

```
h = 1/n;
dt = 0.9*h*h*0.5;
M = round(0.2/dt)
V = 10;
r = dt/(h*h);
%
x = [0:n]*h;
uo = sin(pi*x)
%
for j=1:M
    u = uo;
    for k=2:Nm
        udt2 = uo(k-1) - 2*uo(k) + uo(k+1);
        U = 0.5*V*uo(k)*(uo(k+1) - uo(k-1))/h;
        u(k) = uo(k) + r*udt2 + dt*U;
    end
    uo = u;
end
xh = [0.5 0.5];
uh = [0 max(u)];
plot(x,u,'k-',xh,uh,'k:')
```

Fig. 9.2 shows the plot for $V = 10$, $h = 1/16$.

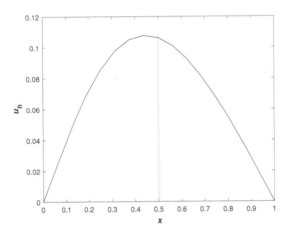

Figure 9.2   Solution to Problem 10 for $V = 10$, $h = 1/16$.

11. Write down the nonlinear system that results from applying the implicit method to the previous problem, using $h = \frac{1}{8}$. How might you try to solve this system?

    Solution: The problem is that the algebraic system is now *nonlinear*, which means we need to use the techniques from §7.11.

12. One way to attack this kind of nonlinear system would be to treat the u_{xx} term as usual in the implicit method, but treat the nonlinear term *explicitly*, i.e., use $(uu_x)(x_i, t_n)$

in the discretization. Write a program to do this approximation and compare your results to the fully explicit method in Problem 10.

◁ • • • ▷

9.2   FINITE ELEMENT METHODS FOR THE DIFFUSION EQUATION

Exercises:

1. Use a finite element approach to solve the problem

$$
\begin{aligned}
u_t &= a u_{xx}, \quad 0 \le x \le 1, \quad t > 0, \\
u(0, t) &= 0, \\
u(1, t) &= 0, \\
u(x, 0) &= u_0(x) = \sin \pi x + \sin 4\pi x.
\end{aligned}
$$

For $a = 1/\pi^2$, this has exact solution $u(x, t) = e^{-t} \sin \pi x + e^{-16t} \sin 4\pi x$. (Confirm this.[1])

2. Use a finite element approach to solve the problem

$$
\begin{aligned}
u_t &= a u_{xx} + 1, \quad 0 \le x \le 1, \quad t > 0, \\
u(0, t) &= 0 \\
u(1, t) &= 0 \\
u(x, 0) &= u_0(x) = \phi_M^h(x),
\end{aligned}
$$

where $\phi_M^h(x)$ is the "hat function" centered nearest the middle of the interval. Solve the problem for a range of (positive) values of a, and compute out to $t = 1$. How does the value of a affect the results?

Solution: The point is that the size of a will affect how fast the initial function smears out and decays away. For "large" a (fast diffusion) this will happen rapidly. For "small" a, it will happen slowly.

3. Discuss how you know your code is working in the above problem, given that we have no exact solution.

4. Use a finite element approach to solve the problem

$$
\begin{aligned}
u_t &= u_{xx}, \quad 0 \le x \le 1, \quad t > 0, \\
u(0, t) &= 1, \\
u(1, t) &= 0, \\
u(x, 0) &= u_0(x) = \sin \pi x.
\end{aligned}
$$

---

[1]The author is chagrined to admit that several exercises in the earlier editions had very "inexact exact solutions." He would like to try to claim it was all done deliberately, in order to catch unwary students who failed to check these things, but that would be a claim of dubious honesty. So, be advised!

For various values of h, solve this using the finite element method. Plot your solutions for several values of t.

5. Repeat the previous problem, this time using different values of a, the diffusion coefficient.

◁ ● ● ● ▷

9.3 DIFFERENCE METHODS FOR POISSON EQUATIONS

Exercises:

1. Prove Theorem 9.4.

   Solution: The key point is that the exact solution and the iterate satisfy almost the same equation. We have

   $$Au = f \leftrightarrow (M - N)u = f \leftrightarrow Mu = Nu + f$$

   and

   $$Mu^{(k+1)} = Nu^{(k)} + f$$

   so that

   $$M\left(u - u^{(k+1)}\right) = N\left(u - u^{(k)}\right),$$

   from which everything follows.

2. Use the sparse matrix construction and solution to solve Example 9.5 with $a = 5$ and $a = 7$, using a sequence of values of h similar to what we used here. Comment on your results. Plot your approximate solution as a surface plot and a contour plot. Do you get the expected convergence rate?

   Solution: You should, for small enough values of h.

3. Set $u(x, y) = x(1 - x)y(1 - y)e^{ax+by}$ for real parameters a and b. Note that $u(x, y) = 0$ on the boundary of the unit square $R = (0, 1) \times (0, 1)$.

   (a) Compute $\frac{\partial^2 u}{\partial x^2}$ and $\frac{\partial^2 u}{\partial y^2}$.

   (b) Define $f(x, y) = -\frac{\partial^2 u}{\partial x^2} - \frac{\partial^2 u}{\partial y^2}$, so that

   $$-\Delta u = f, (x, y) \in R = (0, 1) \times (0, 1),$$
   $$u = 0, (x, y) \in \Gamma = \partial R.$$

   Therefore, u is the solution to a Poisson equation.

   (c) Plot u as both a surface plot and a contour plot, using several different values of a and b.

   Solution: A contour plot for $a = 5$ and $b = 1$ is in Fig. 9.3. Taking larger a and/or b produces steep gradients along the top and/or right boundaries of our domain, which will make approximation more difficult.

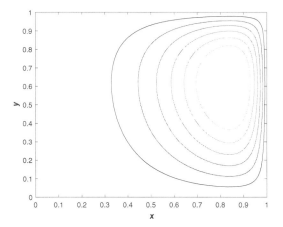

Figure 9.3 Contour plot for Problem 3, using $a = 5$ and $b = 1$.

4. For u and f as defined in the previous problem, write a script that solves the Poisson problem, using the sparse matrix construction and solution, similar to what we did in Examples 9.5 and 9.6. Confirm that the program is working by the decrease in the error as h decreases. Plot the approximate solution and the error as surface plots and contour plots.

Solution: Taking $a = 5$ and $b = 1$, we get the results in Table 9.1. A least-squares fit to the log of the error data yields says that the error is going like h^r for $r = 1.9916 \approx 2$, which is what we expect.

Table 9.1 Results for Problem 4.

N	N_s	T1 (Assembly)	T2 (Solution)	err
8	49	0.0059823000	0.0054801000	$4.10437996 \times 10^{-1}$
12	121	0.0018059000	0.0002722000	$1.95558362 \times 10^{-1}$
16	225	0.0027992000	0.0003592000	$1.12143411 \times 10^{-1}$
24	529	0.0068355000	0.0005886000	$4.99706698 \times 10^{-2}$
32	961	0.0122442000	0.0017577000	$2.82859297 \times 10^{-2}$
48	2209	0.0268254000	0.0046833000	$1.25922921 \times 10^{-2}$
64	3969	0.0392557000	0.0094177000	$7.08726075 \times 10^{-3}$
96	9025	0.0770988000	0.0254033000	$3.15146362 \times 10^{-3}$
128	16129	0.1549921000	0.0419977000	$1.77353169 \times 10^{-3}$
192	36481	0.3113019000	0.1012023000	$7.88417679 \times 10^{-4}$
256	65025	0.5315500000	0.1896888000	$4.43485636 \times 10^{-4}$
384	146689	1.2277448000	0.5206538000	$1.97116626 \times 10^{-4}$
512	261121	2.1909845000	1.0104493000	$1.10878288 \times 10^{-4}$
768	588289	5.2502572000	2.8265900000	$4.92799118 \times 10^{-5}$
1,024	1046529	9.4314992000	5.5223161000	$2.77199657 \times 10^{-5}$
1,536	2356225	21.6673181000	13.7837160000	$1.23199950 \times 10^{-5}$
2,048	4190209	37.3648839000	29.7608316000	$6.92995923 \times 10^{-6}$

5. For the same example as above, use the Collatz high-order scheme to compute your solutions. Again, plot the solution as a surface and as a contour plot.

Solution: Again, using $a = 5$ and $b = 1$ we get the results in Table 9.2. A least squares fit confirmed that the error was $\mathcal{O}(h^r)$ for $r = 3.8287 \approx 4$.

Table 9.2   Results for Problem 5.

N	N_s	T1 (Assembly)	T2 (Solution)	err
8	49	0.0048234000	0.0002067000	$2.41817308 \times 10^{-2}$
12	121	0.0059027000	0.0002257000	$4.84945610 \times 10^{-3}$
16	225	0.0102458000	0.0002910000	$1.57062806 \times 10^{-3}$
24	529	0.0221742000	0.0006693000	$3.11560933 \times 10^{-4}$
32	961	0.0398296000	0.0023147000	$9.85271613 \times 10^{-5}$
48	2209	0.1024933000	0.0023128000	$1.95240394 \times 10^{-5}$
64	3969	0.1595073000	0.0093088000	$6.18279416 \times 10^{-6}$
96	9025	0.3696212000	0.0216570000	$1.22127469 \times 10^{-6}$
128	16129	0.6527077000	0.0476293000	$3.86515991 \times 10^{-7}$
192	36481	1.5160478000	0.1340260000	$7.63524937 \times 10^{-8}$
256	65025	2.7006973000	0.3384284000	$2.41592848 \times 10^{-8}$
384	146689	6.1828549000	0.9409689000	$4.77355622 \times 10^{-9}$
512	261121	10.4713138000	1.5550072000	$1.51248125 \times 10^{-9}$
768	588289	24.4528745000	4.4541261000	$3.03909342 \times 10^{-10}$
1,024	1046529	42.8830576000	9.9697071000	$1.04506181 \times 10^{-10}$
1,536	2356225	103.8321026000	33.0659032000	$4.27080593 \times 10^{-11}$
2,048	4190209	197.0060145000	75.5076653000	$5.01989561 \times 10^{-11}$

6. Show that the Jacobi iteration (9.43) is equivalent to the matrix iteration (7.19) from Chapter 7.

7. Show that the Gauss-Seidel iteration (9.44) is equivalent to the matrix iteration (7.20) from Chapter 7.

8. Show that the SOR iteration (9.45) is equivalent to the matrix iteration (7.22) from Chapter 7.

9. What is the truncation error in the approximation defined by (9.39)?

   Solution: Each derivative approximation is $\mathcal{O}(h^2)$ so the truncation error is $\mathcal{O}(h^2)$.

10. Let $x_1, x_2, \ldots, x_{n-1}, x_n$ be orthogonal vectors in a vector space V of dimension n. Show that if $z \in V$ is orthogonal to each one of the x_k, then $z = 0$.

    Solution: Let x be an arbitrary vector, and expand it in terms of the x_j as follows:

    $$x = \sum_{j=1}^{n} \xi_j x_j$$

    Then

    $$(z, x) = \sum_{j=1}^{n} \xi_j (z, x_j) = 0.$$

    But the only vector that is orthogonal to all other vectors is the zero vector.

11. For an iteration of the form

    $$u^{(k+1)} = Tu^{(k)} + c,$$

show that

$$\frac{\|u^{(k+1)} - u^{(k)}\|_\infty}{\|u^{(k)} - u^{(k-1)}\|_\infty} \leq \|T\|_\infty.$$

Can we use this to estimate ρ_J and therefore ω_*? Hint: Recall Exercise 8 from Section 7.7.

Solution: We have that

$$u^{(k+1)} = Tu^{(k)} + c$$

and

$$u^{(k)} = Tu^{(k-1)} + c,$$

therefore,

$$u^{(k+1)} - u^{(k)} = T(u^{(k)} - u^{(k-1)}).$$

Hence,

$$\|u^{(k+1)} - u^{(k)}\|_\infty \leq \|T\|_\infty \|(u^{(k)} - u^{(k-1)})\|_\infty,$$

or,

$$\frac{\|u^{(k+1)} - u^{(k)}\|_\infty}{\|u^{(k)} - u^{(k-1)}\|_\infty} \leq \|T\|_\infty.$$

We know that $\|T\| \geq \rho(T)$; therefore, we use this as an estimate for ρ_J by running a few iterations of the Jacobi method, then using this in the formula for ω_*.

12. For the same family of problems as in Problem 2, write a script that does the SOR iteration to compute a solution, using a convergence criterion of $\tau = 10^{-8}$.

13. Find the optimal values of ω for SOR when using the Collatz high-order difference method.

14. Use your "Collatz optimal ω" values in an SOR iteration to produce solutions to the examples from Problem 5.

15. In both §6.10.2 and §7.10, we attacked an iteration process by doing a few initial iterations on a very coarse grid, then mapping that crude solution to a finer grid to get our final solution. Apply this idea to our SOR iteration for Example 9.6; first, run sample calculations on your computer to find out how long it takes your platform to converge for $N = 1024$, using the convergence criterion of $\tau = 10^{-10}$; next, run experiments using different initial meshes and number of iterations at each coarse mesh, and try to get a final solution in significantly less time than when just doing the naive iteration.

16. Discretize the PDE

$$\begin{aligned}
-u_{xx} - u_{yy} + bu_x &= f; & (x,y) &\in (0,1) \times (0,1); \\
u(x,0) = u(x,1) &= 0; & x &\in (0,1); \\
u(0,y) = u(1,y) &= 0; & y &\in (0,1);
\end{aligned}$$

in the case $h = \frac{1}{3}$, $b \neq 0$. Use an ordinary central difference approximation for the u_x term. Is the resulting system symmetric?

Solution: No, it is not.

17. Consider the linear system problem

$$Au = f,$$

where A is *not* symmetric. Modify the CG algorithm to solve the symmetrized system

$$A^T Au = A^T f$$

without explicitly forming the matrix $A^T A$.

18. Write a code to implement the algorithm you wrote in Problem 17. Test it on the system obtained by discretizing the PDE

$$
\begin{aligned}
-u_{xx} - u_{yy} + u_x &= \pi \cos \pi x \cos \pi y + 2\pi^2 \sin \pi x \sin \pi y; \ (x,y) \in (0,1) \times (0,1); \\
u(x,0) = u(x,1) &= 0; x \in (0,1); \\
u(0,y) = u(1,y) &= 0; y \in (0,1);
\end{aligned}
$$

for $h = \frac{1}{4}$ and $h = \frac{1}{8}$. The exact solution is $u(x,y) = \sin \pi x \sin \pi y$; use this to ensure your algorithm is working properly.

19. How does the discretization change when the boundary data is nonhomogeneous (i.e., nonzero)? Demonstrate by writing down the discrete system for the PDE

$$
\begin{aligned}
-u_{xx} - u_{yy} &= -2e^{x-y}; (x,y) \in (0,1) \times (0,1); \\
u(x,0) = e^x; \quad u(x,1) &= e^{x-1}; \quad x \in (0,1); \\
u(0,y) = e^{-y}; \quad u(1,y) &= e^{1-y}; \quad y \in (0,1);
\end{aligned}
$$

for $h = \frac{1}{4}$. *Hint:* It will help to write the values of the approximate solution at the grid points in two vectors, one for the interior grid points, where the approximation is unknown, and one at the boundary grid points, where the solution is known.

20. Apply the following solution techniques to the system in Problem 19, this time using $h = \frac{1}{8}$. Use the exact solution of $u(x,y) = e^{x-y}$ to verify that the code is working properly. (Confirm, first, that this is indeed the exact solution.)

    (a) Direct solution using the sparse matrix construction;

    (b) Jacobi iteration;

    (c) Gauss–Seidel iteration;

    (d) SOR iteration using $\omega = 1.4465$;

    (e) Conjugate gradient iteration.

    Use a sequence of grids to confirm that the error is $\mathcal{O}(h^2)$.

◁ • • • ▷

CHAPTER 10

AN INTRODUCTION TO SPECTRAL METHODS

10.1 SPECTRAL METHODS FOR TWO-POINT BOUNDARY VALUE PROBLEMS

Exercises:

1. Write a program to solve the BVP

$$
\begin{aligned}
-u'' + 5u' + u &= 1, \quad -1 \leq x \leq 1; \\
u(-1) &= 0; \\
u(1) &= 1;
\end{aligned}
$$

using the spectral method, with either "boundary bordering" or "basis recombination." The exact solution is

$$u(x) = Ae^{r_1 x} + Be^{r_2 x} + 1,$$

for $r_1 = 5.19258$, $r_2 = -0.19258$, and

$$A = 0.003781 \quad B = -0.824844$$

(confirm this); plot the solution and the error, and produce a table of maximum absolute errors, for $4 \leq N \leq 32$.

Solutions Manual to Accompany An Introduction to Numerical Methods and Analysis, Third Edition.
James F. Epperson.
© 2021 John Wiley & Sons, Inc. Published 2021 by John Wiley & Sons, Inc.

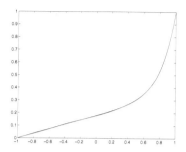

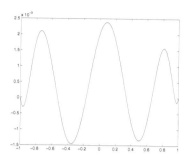

Figure 10.1 Solution to Problem 1, $N = 8$.

Figure 10.2 Error for Problem 1, $N = 8$.

Solution: The author got the solution and error plots in Fig. 10.1 and Fig. 10.2 for $N = 8$.

2. Solve the BVP in Example 10.2, for $4 \leq N \leq 32$, but take advantage of the fact that the solution is even by looking for an approximation that uses only the even Chebyshev polynomials.

3. It has been suggested that a better way to do "basis recombination" would be as follows:

$$
\begin{aligned}
C_{2N}(x) &= C_{2N}(x) - C_{2N-2}(x), \\
C_{2N-1}(x) &= C_{2N-1}(x) - C_{2N-3}(x).
\end{aligned}
$$

Repeat Problem 1 using this basis. In addition to the same plots as requested in Problem 1, plot the condition number of the matrix A as a function of N for both methods.

4. Extend the work in Theorem 10.1 to include third and fourth derivatives.

Solution: This is a simple exercise in the calculus. You should get

$$
\frac{d^3 T_n}{dx^3} = \frac{-n^3 \sin^2 t \sin nt - 3n^2 \cos t \sin t \cos nt + (3\cos^2 t + \sin^2 t)n \sin nt}{\sin^5 t},
$$

and

$$
\frac{d^4 T_n}{dx^4} = \frac{n^4 \sin^3 t \cos nt - 6n^3 \cos t \sin^2 t \sin nt - n^2 A \cos nt + nB \sin nt}{\sin^7 t},
$$

where

$$
A = (15 \cos^2 t \sin t + 4 \sin^3 t),
$$

and

$$
B = (9 \cos t \sin^2 t + 15 \cos^3 t).
$$

These formulas are taken from p. 325 of Boyd's book [3].

5. Use the formulas from the previous problem to approximate the solution to

$$u'''' = 1,$$
$$u(-1) = u(1) = 0,$$
$$u'(-1) = u'(1) = 0.$$

Compute a spectral approximation for $4 \leq N \leq 32$. Plot your solution for $N = 16$.

6. Compute the condition number of the matrices in the previous exercise, as a function of N.

7. As an alternative to the trigonometric formulas from Theorem 10.1, we could use the three-term recursion (10.2) as a basis for constructing the spectral coefficient matrix. Show that

$$T'_{n+1}(x) = 2T_n(x) + 2xT'_n(x) - T'_{n-1}(x), \quad T'_0(x) = 0, \quad T'_1(x) = 1,$$

and similarly for the second derivative. Write a program to solve Problem 1 in this way. Use MATLAB's `tic-toc` timing commands to compare the costs of forming the spectral coefficient matrix this way, compared to the procedure outlined in the text.

◁ ● ● ▷

10.2 SPECTRAL METHODS IN TWO DIMENSIONS

Exercises:

1. Write your own script to use the spectral method, as outlined here, to solve Example 10.6 using $a = 7$ and $a = 9$.

2. Set $u(x, y) = (1 - x^2)(1 - y^2)e^{ax+by}$ for real parameters a and b. Note that $u(x, y) = 0$ on the boundary of the square $R = (-1, 1) \times (-1, 1)$.

   (a) Compute $\frac{\partial^2 u}{\partial x^2}$ and $\frac{\partial^2 u}{\partial y^2}$.

   (b) Define $f(x, y) = -\frac{\partial^2 u}{\partial x^2} - \frac{\partial^2 u}{\partial y^2}$, so that

   $$-\Delta u = f, (x, y) \in R = (-1, 1) \times (-1, 1),$$
   $$u = 0, (x, y) \in \Gamma = \partial R.$$

   Therefore, u is the solution to a Poisson equation.

   (c) Plot u as both a surface plot and a contour plot, using several different values of a and b.

   Solution: This is essentially Problem 3 from §9.3, adapted to the spectral setting, which prefers a domain $R = (-1, 1) \times (-1, 1)$. Taking $a = -4$ and $b = 8$ yields the contour plot in Fig. 10.3.

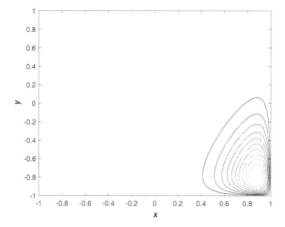

Figure 10.3 Contour plot for Problem 2, using $a = -4$ and $b = 8$.

The contour plot suggests that the solution is very steep near the boundary, which will challenge even the spectral method.

3. Use your spectral solution script to solve the problem constructed in the previous exercise, for a variety of a and b values.

    Solution: Taking $a = -4$ and $b = 8$, we get the results summarized in Fig. 10.4, Fig. 10.5, and Table 10.1.

    Note that the error for the smaller values of N is actually quite large—about 2,400, for $N = 4$—which is not surprising given the contour plot of the exact solution. The error does decline rapidly as we take for terms, but we have to go all the way out to $N = 28$ before the error begins to saturate. The coefficient decay plot, Fig. 10.5, does suggest we have converged, in the sense that taking more terms will not decrease the error significantly.

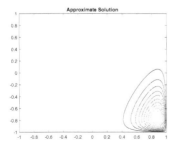

Figure 10.4 Contour plot of approximate solution for $N = 43$; compare to Fig. 10.3, above.

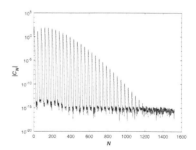

Figure 10.5 Coefficient decay for $N = 43$.

Table 10.1 Spectral solution results for Problem 10.2.3.

n	N	N_e	$T_1 =$ Assembly time	$T_2 =$ Solution time	Error
4	9	9	0.0055498000	0.0001113000	2.41165488×10^3
5	16	16	0.0028922000	0.0000979000	1.18468209×10^3
6	25	25	0.0026896000	0.0002161000	8.47824909×10^2
7	36	36	0.0014407000	0.0001884000	5.15084348×10^2
8	49	49	0.0054764000	0.0005592000	2.35532078×10^2
9	64	64	0.0030302000	0.0002058000	91.0403906
10	81	81	0.0042291000	0.0002683000	31.3853341
11	100	100	0.0064572000	0.0003052000	10.4762331
12	121	121	0.0093239000	0.0004165000	3.51010979
13	144	144	0.0132698000	0.0006441000	1.07009551
14	169	169	0.0174758000	0.0011743000	$2.96428388 \times 10^{-1}$
15	196	196	0.0243783000	0.0008237000	$7.82311112 \times 10^{-2}$
16	225	225	0.0334466000	0.0017701000	$1.98394961 \times 10^{-2}$
17	256	256	0.0527671000	0.0023140000	$4.72073947 \times 10^{-3}$
18	289	289	0.0596898000	0.0023163000	$1.04475757 \times 10^{-3}$
19	324	324	0.0702664000	0.0026417000	$2.21789029 \times 10^{-4}$
20	361	361	0.0877788000	0.0036661000	$4.48400134 \times 10^{-5}$
21	400	398	0.1005163000	0.0033801000	$8.67520623 \times 10^{-6}$
22	441	418	0.1259192000	0.0049139000	$1.58912536 \times 10^{-6}$
23	484	412	0.1508022000	0.0053836000	$2.78711298 \times 10^{-7}$
24	529	415	0.1708327000	0.0064900000	$4.75048836 \times 10^{-8}$
25	576	425	0.2109273000	0.0080198000	$7.69495045 \times 10^{-9}$
26	625	434	0.2489746000	0.0109425000	$1.20326149 \times 10^{-9}$
27	676	442	0.2802686000	0.0140222000	$1.85963245 \times 10^{-10}$
28	729	446	0.3602121000	0.0230818000	$3.64934749 \times 10^{-11}$
29	784	450	0.3933945000	0.0256014000	$3.41491041 \times 10^{-11}$
30	841	450	0.4528874000	0.0345000000	$1.20727248 \times 10^{-11}$
31	900	449	0.4967433000	0.0326247000	$2.05853321 \times 10^{-11}$
32	961	449	0.5619765000	0.0557487000	$1.85778198 \times 10^{-11}$
33	1,024	450	0.6444865000	0.0687930000	$2.88902235 \times 10^{-11}$
34	1,089	450	0.7409033000	0.0694104000	$1.81472615 \times 10^{-11}$
35	1,156	449	0.8851444000	0.0633313000	$4.67821337 \times 10^{-11}$
36	1,225	450	0.9990630000	0.0950365000	$2.13731255 \times 10^{-11}$
37	1,296	450	1.1342848000	0.0960442000	$2.80340473 \times 10^{-11}$
38	1,369	450	1.2022592000	0.1495144000	$1.37058143 \times 10^{-11}$
39	1,444	449	1.3825504000	0.1324822000	$3.25073787 \times 10^{-11}$
40	1,521	450	1.4657469000	0.1623935000	$2.54942734 \times 10^{-11}$
41	1,600	450	1.7576274000	0.1934468000	$4.89706053 \times 10^{-11}$
42	1,681	450	1.8699778000	0.2081700000	$3.11846104 \times 10^{-11}$
43	1,764	450	2.1324582000	0.2636492000	$2.29647412 \times 10^{-11}$

◁ ● ● ▷

10.3   SPECTRAL METHODS FOR TIME-DEPENDENT PROBLEMS

Exercises:

1. Use spectral collocation with Crank-Nicolson time-stepping to solve the following PDE:

$$
\begin{aligned}
u_t &= u_{xx}, \\
u(-1, t) &= 0, \\
u(1, t) &= 0, \\
u(x, 0) &= \cos \pi x / 2 - \sin 4\pi x.
\end{aligned}
$$

The exact solution is $u(x, t) = e^{-\pi^2 t/4} \cos \pi x/2 - e^{-16\pi^2 t} \sin 4\pi x$. Compute out to $t = 1$; use a sequence of values of N; plot your approximation and the error for one of them at $t = 1$.

2. Use your spectral code to solve the problem

$$
\begin{aligned}
u_t &= a u_{xx}, \\
u(-1, t) &= 1, \\
u(1, t) &= 1, \\
u(x, 0) &= (x^2 - 1)^8.
\end{aligned}
$$

Assume $a = 1$ and compute out to $t = 1$, using a sequence of values of N. Plot the solution profile as the computation advances. Now vary a (you must keep it positive, of course) and investigate how this affects the solution.

Solution: The initial condition approximates a "pulse" in the center of the interval. As the system evolves, the pulse will spread out and decay. Higher values of a (the diffusion coefficient) will speed up this process.

3. Now change the initial condition to $u(x, 0) = (x^4 - 1)^8$ and repeat the above problem.

Solution: This is essentially the same as the previous problem, but with a more sharply defined "pulse."

4. Consider how to implement spectral collocation with variable coefficients. Construct the general linear system that would result from solving the problem:

$$
\begin{aligned}
u_t &= a(x) u_{xx}, \\
u(-1, t) &= 0, \\
u(1, t) &= 0, \\
u(x, 0) &= u_0(x).
\end{aligned}
$$

5. Apply your results from the above problem to approximate solutions to

$$
\begin{aligned}
u_t &= a(x) u_{xx}, \\
u(-1, t) &= 0, \\
u(1, t) &= 0, \\
u(x, 0) &= \cos \pi x / 2,
\end{aligned}
$$

for the following choices of a:

(a) $a(x) = (1 + x^2)$;

(b) $a(x) = e^{-x^2}$;

(c) $a(x) = (1 - x^2)$ (because a vanishes at the boundary, this problem is known as *degenerate* but you should be able to compute solutions).

6. Consider the nonlinear problem

$$
\begin{aligned}
u_t &= u_{xx} + Vuu_x, \\
u(-1, t) &= 0, \\
u(1, t) &= 0, \\
u_0(x) &= \sin \pi x.
\end{aligned}
$$

Use spectral collocation to attack this problem as suggested in Problem 12 of §9.1, by treating the nonlinearity explicitly (at time $t = t_n$) and the differential equation implicitly (at time $t = t_{n+1}$). Compare your spectral solution to the explicit solution computed in Problem 10 of §9.1. Comment on your results.

◁ • • • ▷

10.4   CLENSHAW-CURTIS QUADRATURE

Exercises:

1. Use the appropriate change of variable to show how to apply Clenshaw-Curtis quadrature to an integral over an arbitrary interval $[a, b]$.

   Solution: The change of variable is the same as obtained in §5.6 for Gaussian quadrature:

   $$
   \int_a^b g(x)\,dx = \int_{-1}^1 f(z)\,dz,
   $$

   for

   $$
   f(z) = \frac{1}{2}(b - a)g\left(a + \frac{1}{2}(b - a)(z + 1)\right).
   $$

2. Write a program to do Clenshaw-Curtis quadrature on each of the integrals in Problem 4 of §5.6. Compare your results to those obtained with Gaussian quadrature. Produce a log-log plot of the error as a function of N for each integral.

   Solution: We will do (b) and (c). The results are summarized in Table 10.2.

3. Looking at the plots in Fig. 10.24, we see that most of them show a very rapid decrease of the error, and a "rounding error plateau" is reached for most of the examples. The exception is the last one, where the integrand is given by $f(x) = \sqrt{1 - x^2}$. Explain why this example is the one that displays this kind of sub-optimal performance.

   Solution: The integrand is singular at the endpoints of the interval (the derivatives blow up) so this slows the convergence.

Table 10.2   Solution for 10.3.2(b) and (c).

N	10.3.2(b)	Error	10.3.2(c)	Error
2	0.65367552526213	−0.00702835762825	1.32175583998232	−0.02591897397799
4	0.64425803338082	0.00238913425305	1.29638987827887	−0.00055301227455
8	0.64664682519792	0.00000034243595	1.29583765211218	−0.00000078610785
16	0.64664716763387	0.00000000000000	1.29583686600986	−0.00000000000553
32	0.64664716763387	0.00000000000000	1.29583686600433	0.00000000000000

4. Let $C_N = \sum_{k=1}^{N} w_k^{(N)} f(\xi_k^{(N)}) \approx \int_{-1}^{1} f(x)dx$ be the Clenshaw-Curtis quadrature operator. Show that C_{2N} uses some of the same function values as C_N. Why is this important?

◁ ● ● ● ▷

Printed and bound by CPI Group (UK) Ltd, Croydon, CR0 4YY